Come Back to Me

A Story of Dying and Everything After

By Jay Little

I died on Tuesday afternoon, May 24th, 2011

I had just suffered sudden cardiac death. Yet somehow, I was revived, shocked back to life. Afterwards, a combination of amazing doctors, nurses, and technology worked on restoring my physical body. Even after resuscitation, I was in critical condition, hanging on by a thread. Meanwhile, my mind wandered aimlessly through the dark corners of oblivion for weeks. Against all odds, I survived. My physical recovery was called "miraculous" and amazed everyone. But my emotional, mental, and spiritual recovery took much, much longer. Looking back on it now, I'm convinced my survival all started with a kiss and a whisper … *come back to me.*

This is my story. It is a story about a 38-year old man who suffered a massive heart attack – my second in eight months – followed by a number of serious, life-threatening complications. It is a story of the frightening obstacles and powerful miracles encountered along my path to recovery, using a unique blend of perspectives. Throughout my recovery, my wife kept a detailed journal chronicling the objective aspects, such as specific dates, medical issues, and milestones. Kept under sedation for several weeks, and still groggy from the medication and physical ordeal for months afterward, I provide a very different, subjective look at my recovery – how I was feeling, what I experienced, and what I've learned from these events.

At times, writing this book allowed me to look back and find the silver linings and even moments of humor along the way. More importantly, this book forced me to acknowledge and accept these events so that catharsis could do its job. That process of coping and recovering is much easier to do now that the worst of the danger has passed and life has returned to as normal as it will ever be again.

Thank you to Trish, my wife, for calling me back. Thank you to Dr. Cindy Martin for "turning up the volume" so I could hear her calling. Thank you to the countless other nameless people I never even met, but who were involved in my complex, convoluted recovery process. Thank you to all the people who encouraged me to write this book – it was an incredibly cathartic experience. And finally, thank you to all our friends and family who stopped their lives momentarily so we could move forward with ours.

As with everything else in life, individual cooking times, mileage, and weight loss may vary. This person's results are definitely not typical.

Cover: The Orion Nebula
Courtesy of NASA, ESA, and The Hubble Heritage Team (STScI/AURA)

Come Back to Me © 2012 Jason Patrick Little. All Rights Reserved.
ISBN 978-1-105-48350-9

Come Back to Me was written from August 2011 through April 2012. References to any registered trademarks or copywritten material do not imply anything, challenge anything, and were not used with anyone's consent. Not all persons mentioned herein are fictional and not all references are coincidental. But a great many of them are.

Come Back to Me

Table of Contents

Foreword by Trish Little ....
1. The Beginning of an End 1
2. The CaringBridge Connection 4
3. May 8
4. June 16
5. July 39
6. August and Onward 54
7. On Nurses & Doctors 60
8. So What's With the Title? 64
9. Jedi Mind Tricks 66
10. Facing Fear 91
11. Taken for Granted 94
12. Dignity is a Lie 98
13. If I Hear "You're Too Young For This" One More Time ... 99
14. Tools of Recovery 104
15. For the Record 112
16. The Idiot Box 126
17. The Discouragement Center 130
18. Introspection 138
19. By the Numbers 145
20. Are We There Yet? 151
21. Odds & Ends 154
22. The Big Finish 160
Additional Thanks 168
About the Author 169

A note on medical terminology in this book

Trish and I are not medical experts by any stretch of the imagination, although we now know a whole lot more about certain medical phenomena than most people would ever want to.

Throughout this book I wouldn't be surprised if you see some medical terms used incorrectly or inconsistently. For example, I often use coma, comatose, unconscious and sedated interchangeably. Regardless of the exact physiological definition of my condition at the time, they all relate to me being unresponsive, immobile, and completely unaware of my surroundings.

It's also possible I misinterpreted some of the dense technical language from my medical records despite my research. Regardless, the book is based on my initial reactions and the feelings evoked after reading these records and upon hearing the accounts of the event from medical staff and family, as well as processing the information I learned as I conducted more research for this book. Hopefully the context of the various references will make it clear enough to follow along.

Foreword by Trish Little

"Mrs. Little, I wanted to let you know that we've called a Code Blue for your husband."

  Those might not have been the exact words of the doctor whose name vanished from my mind almost instantly, but that's how I remember it. I was in an elevator with random strangers, on my way to keep my husband company in the hospital. The doors had just closed after letting someone out on the second floor. Then my cell phone rang. I was expecting it to be Jay, asking what was taking me so long.

  I was actually a little irked at him. He had just spent the previous weekend in St. Louis with friends, and had been late in returning on Sunday. Jay has always been directionally-challenged, and when his GPS sent him the long way home, he followed it, adding about two hours to his nine hour drive back.

  Monday afternoon Jay called to tell me he was feeling some symptoms (chest pain, etc.) similar to those he felt before his heart attack in September 2010. He decided to drive himself to the hospital to get checked out. I didn't panic. I actually saw it as another aggravation that would probably turn out just fine. I asked if he wanted me to meet him there, but he said no. It wasn't a big deal.

  Jay called me later that evening, saying that they had stopped the pain and he was feeling better, but they wanted to keep him overnight to be sure. I offered to come visit him with our two sons, but again, it didn't seem like that big a deal. After all, Jay would be home tomorrow anyway.

  Tuesday I had several conversations at work along the lines of, "Why are you at work? You should be with Jay." My response was, "He said not to worry about it, so I'm not worrying about it." He called me after they had given him a stress test. He sounded exhausted, but that is kind of the point of the test.

  Then around noon I was talking to a co-worker at her desk, and heard my phone ring but didn't run to answer it. When I checked the voice mail, it was from Jay.

  "As soon as you get this, come see me."

  He still sounded tired, but something else too – sad, or lonely, or scared. So while my computer was shutting down I told my manager that I would be out the rest of the afternoon. Then I called to tell Jay I'd be there as soon as I could. I drove, I parked, I got in the elevator. Then I got the call.

And the rest, as they say, is history.

Calling our families. Sitting in waiting rooms. Hoping for the best. Fearing the worst. Sobbing uncontrollably. Staring vacantly into space, unable to process a coherent thought. The social worker gently reminding me that life is still going on, prodding me to think of who else I need to call, who was going to pick up the kids from school and daycare – the boring logistics of day to day life that can actually help keep you grounded when you feel like the earth has suddenly given way under your feet.

I've had many people tell me how "strong" I was during all of this. I still don't know what that really means. If I would have thought that lying in bed in the fetal position was actually a valid option that would have helped anything, I probably would have done that. But I discovered that trying to keep our children as close to their regular routine as possible helped me, too. Well, having my parents or Jay's parents stay with us pretty much the whole summer helped a lot, too. For conversation, distraction, and housekeeping.

So here we are, almost a year later. Making some positive changes while trying to keep from slipping back into some of our previous routines and bad habits. Trying to live our lives like most people, acting as if the ground is solid under our feet, and we have all the time in the world. Trying to forget that the ground can fall away suddenly and without warning. Pretending sometimes that we weren't as close as we actually were to never getting that ground back again.

Are things back to normal? They never will be. We just have to get used to this new type of normal, even if it does seem to keep shifting around beneath us. In the end, I keep coming back to my answer to the chaplain who accompanied me to the parking garage at the hospital while Jay was being transferred to the intensive care unit. Since I was still recovering from the emotional trauma of all this, he asked me if I was sure that I was okay to drive.

"I will be."

1. The Beginning of an End

Monday, May 23rd, 2011, started out as a great day. I love my job – I am a professional game designer and get to work on a wide variety of board games, card games and roleplaying games. I had recently received a promotion, and was now managing the game design team. Things had gone so well that several more team members were assigned to report to me. And I still got to work on some very exciting projects of my own. We hadn't officially announced it yet, but Fantasy Flight Games – the company I work for – acquired the rights to publish games based on the Star Wars license, and I was in the middle of designing a game about dog-fighting space duels between X-Wings and TIE Fighters – how awesome is that?

Life was great. It is hard to believe just how fortunate we have been over the years. I was married to an amazing, loving wife, and we were about to celebrate our 16th anniversary. We had two wonderful, healthy boys. We lived in a great community, found a home we both love, both had great jobs, and things were on an upswing since a small scare the previous fall.

Back in September 2010, I had a minor heart attack, which was a real wake up call. Thankfully I was only in the hospital for a short time. I was hospitalized just long enough for a series of angioplasty procedures to insert three stents in my coronary vessels. Since then, I had made some significant changes to my lifestyle and diet; I started exercising more and slowly began to lose some weight. In many ways, I felt like I was the healthiest I had been in the previous ten years. Things were going so well in our lives, we certainly didn't expect any more complications.

That Monday afternoon, my boss and I were having a meeting to discuss some upcoming projects and how best to manage the team I was now responsible for. While we were talking, I felt a wave of dizziness and cold wash over me. I suddenly started to sweat profusely. This was quickly followed by a sharp pain in my left armpit, and I immediately grew concerned. After all, these were the same symptoms I had experienced in September when I had my heart attack. But it couldn't possibly be another heart attack, could it? Then another wave of sweating broke over me and the pain in my left armpit started to slowly spread down my arm and across my chest.

I interrupted the meeting and told my boss I had to leave. Immediately. But by the time I had gotten my things and headed out to the car, the sweating had subsided and the pain slightly diminished. Perhaps foolishly, I got in my car and started to drive home instead of calling 911. I thought I was just over-reacting. Everything was probably fine. I just needed to lie down for a bit and get some rest.

On the way home, I hit the evening commute traffic. And everything around me ground down to a complete stop.

Once I was in gridlock, the sweating returned. That was followed by more waves of hot and cold flashes. The pain in my armpit intensified and started reaching all the way down my arm to my hand and fingers. My chest hurt each time I took a breath. I panicked, stuck in the middle of five lanes of traffic backed up along one of the busiest stretches of highway in the Twin Cities during rush hour. I called my wife, unsure if I should try to press on straight to the hospital, or call 911 and hope they could somehow reach me in the midst of the traffic jam.

Not knowing what else to do, I decided to head straight for the emergency room at Methodist Hospital in St. Louis Park. Theoretically, it was not that far from where I was stuck in traffic. At least, not that far as the crow flies. I had no idea how much time to expect with traffic backed up the way it was.

That is when the first miracle happened. Once I had made up my mind to head to the emergency room, traffic suddenly opened up. The traffic did not just ease up. It seemed like the lanes of cars parted around me as I drove. I was able to merge quickly from the onramp and easily get over into the lane I needed. Despite the heavy construction going on around the hospital, I do not think I hit a single red light the entire way there. Miracle two soon followed as I pulled into the ever-swamped parking ramp. At this time of day, the parking lot was completely packed. But just then, a car pulled out of the second spot past the handicapped parking. Perfect timing.

Thank God for such a close spot, because I was starting to feel light-headed. As the waves of hot and cold washed over me, I stumbled into the hospital. I know I would not have made it there if I had parked further up the ramp. I staggered to the information desk. The lady at the desk asked if I was ok. I told her I needed to get to the ER right away, and promptly collapsed into one of the wheelchairs by the desk.

The next few hours are extremely hazy. I remember arriving in the ER, having someone push a clipboard in front of me with some forms, and signing some papers. I remember getting wheeled back to a room and put into a bed. I had some blood drawn and an IV put in. They checked my troponin levels – a protein measured to help diagnose cardiac activity. The first round of labs all came back normal. I was given a dose of nitroglycerin to help manage the pain.

I am pretty sure I was forced to sit up so a technician could put a board behind my back for some x-rays. They drew more blood to check my troponin levels again – and again, everything was good. Then I was given another dose of nitro. Other than that, I'm not entirely sure what took place. And I don't know in what order any of this happened. Just that there were a lot of people buzzing around me.

According to my wife, I was kept overnight for observation, but all the lab work and test results came back fine. Tuesday morning, May 24th, I was given a stress test, where the doctor had me exert myself to raise my blood pressure and heart rate. Just to make sure everything was ok. I vaguely remember being prompted to push harder and go for just a few minutes longer… at something. I do not remember, however, exactly what the stress test was. I only remember being utterly exhausted.

The stress test turned out fine. I was cleared to leave. Everything was completely normal. Thank goodness. I had just been paranoid, and could finally go home. I gathered my things and was about to head out.

That is when everything fell apart. Fast. Apparently I called my wife and told her to come see me immediately, that something was terribly wrong. I do not remember doing that. Weeks later, some stray memories started to trickle back. It is hard to believe that a few vague impressions of the stress test would be the last thing I could remember for nearly a month. But that is how it all began.

With everything being just fine.

2. The CaringBridge Connection

Comments on Comments

The next part of this account is taken directly from Trish's regular updates from *CaringBridge.com*, an incredible free resource for families to manage communication surrounding a tragedy. Each entry is shown as it was listed on the site, which is why some of the accounts and references seem aimed at a different audience. In some spots, I interject my own personal comments and feedback, which are blockquoted in **Arial** to easily distinguish them from the main commentary.

To read the original CaringBridge.com story without interruption, simply skip the indented text. Since this was originally written for the web, I've retained the hard paragraph breaks originally used rather than indented paragraphs found in the other chapters of this book. The original errors and typos are preserved except where noted, to mirror the online content as closely as possible.

Welcome to our CaringBridge website. I've set it up to keep everyone informed because inevitably I'll forget to contact one point in our web of friends and family when there's news to share. Also, just wanted to get this all in one spot so years and years from now we can look back in awe and wonder at how it all worked out just fine. :)

> Which, amazingly, we are already able to start doing.

Background Story

OK, so where to start?

On Monday, [May 23] Jay started feeling unwell at work - numb left arm, palpitations, nausea, I'm not sure what else. He drove himself to Methodist Hospital in St. Louis Park, MN. They gave him meds and he was feeling better, but kept him overnight for observation.

Tuesday morning, he had a stress test. I talked to him afterwards; he sounded tired, but that's kind of the point of the test. Around 12:15 he left me a message at work to come see him as soon as I got the message. I did.

As I was riding in the elevator between 2nd and 3rd floor (he was on 3rd) I got a call from a doctor - they had called a Code Blue and were working to resuscitate him.

Apparently the doctor had gone in to tell him the stress test came out fine....and then he was very not-fine to say the least.

> It wasn't until almost six weeks after my heart attack (maybe three weeks after regaining consciousness?) that I even found out I had suffered a heart attack. And even then it was only in the most vague, high level terms. I didn't know I had flat-lined, or that I had to be shocked back, or most frighteningly of all, that before all of this drama I had been cleared for release and was about to go home.

They got him stable enough to go to the Cath Lab, the Cardiologist was able to clear the clot, but there was a section of his heart that just wasn't pumping very well. They decided to transfer him to U of MN Hospital where they could offer more treatment than Methodist could.

During this time I was never alone - between Ken the chaplain, and Julie the social worker (usually both) I was being well-cared-for at the same time Jay was.

Once they got him on the move to the U, I drove home to pick up the kids and bring them home. Eternal thanks to JR for being able to drop everything and come over to keep an eye on them, and for his patience in having Nate show him 2 minutes of each game in our Wii collection.

Mom and Dad arrived at the house around 7ish, fed the kids and got them to bed. Then Mom began her self-guided tour of Minneapolis and the surrounding area, but you'll have to ask her about that. Suffice to say that U of MN Hospital has TWO campuses, one on each side of the Mississippi. And it's not just Mom that made the mistake - I originally found myself in the Children's Hospital ER on the wrong side of the river.

So during this time, Jay's oxygen level got really low, his lungs had fluid in them, and the doctors at the U decided the best course was to hook him up to an ECMO machine. You can google that, but my understanding is that it helps pump the blood so his heart doesn't have to work so hard, and it was also oxygenating his blood to help out his lungs.

> Straight from Wikipedia: *In intensive care medicine, extracorporeal membrane oxygenation (ECMO) is an extracorporeal technique of providing both cardiac and respiratory support oxygen to patients whose heart and lungs are so severely diseased or damaged that they can no longer serve their function. Initial cannulation of a patient receiving ECMO is performed by a surgeon and maintenance of the patient is the responsibility of the ECMO Specialist and gives 24/7 monitoring care during the duration of the ECMO treatment.*
>
> Only after reading my medical records and seeing my cardiologist almost nine months after my heart attack did I realize what a serious decision this was. The ECMO procedure would be physically taxing and potentially traumatic to both my body and my family. I had been without oxygen for so long, they had to be sure I wasn't already brain dead or beyond the point where it was a viable option. And even then, the success rate for ECMO is frighteningly low – it's a medical Hail Mary Pass.

Mom and I finally got to see him around 11pm, he was in his room in the Cardiac ICU. They were also keeping him hypothermic to protect his body from the low oxygen levels, and sedated and paralyzed to basically let his body do nothing but rest and recover. His heart rhythm had regulated, his labs were all looking good, his kidneys were functioning, so there wasn't much to do at this point but wait. We went home to try to grab what little rest we could find.

Wednesday May 25: You never complain about a day of sitting around in a waiting room doing practically nothing (Mom, Dad and I did complete a 500-piece puzzle in the afternoon) after getting through a day like Tuesday. No news was good news, Jay continued doing well on all the machines he was hooked to. 10pm was the 24-hour mark that they would start warming him up - the cautious time was going to be between 95-97 degrees as that can sometimes throw electrolytes out of whack.

Jay's Mom arrived around 4pm. Given the plan to warm him up over night, and knowing the 'scary' time would be around 3-4am, my folks and I decided to go pick up the kids, get some dinner, try to get some rest and come back to the hospital to spend the night. Judy stayed at the hospital.

Fast forward - we had dinner, kids got to bed eventually, and Mom and I went to get my car out of the Methodist Hospital ramp - I had taken Jay's car for the onboard GPS when they transferred him. Since I'd used my ticket to get his car out, I handed his ticket to get my car out, and the teller asked if maybe I'd given her the wrong ticket? The price displaying was $26. I explained the situation, so she ran it as a "discharge" and only charged me $4. Thank you, anonymous parking teller!

So we got home, packed some bags - snacks, books, cards, pillows, blankets - and Mom and I headed out to the hospital, leaving instructions for Dad to get through the morning with the kids. About 2 blocks away from home, Judy called to say they were delaying warming Jay up until his potassium levels got higher, she was ok to stay by herself, so Mom and I went back home to get better sleep than we could in a very uncomfortable waiting room.

Which brings us to today, Thursday, May 26. And based on the character limit and how few I have left, I'll figure out how to post that as an update.

> I, of course, remember none of this. I was unconscious and kept in a chemically-induced coma so the doctors could try to manage the various things going wrong. Even after I was brought out of sedation a few weeks later, I did not know any of these details. Looking back at things, I'm glad I didn't know. Once I started to come around, I was more and more scared to learn about what had actually happened.
>
> During this time, the initial stress on Trish must have been unbelievable. I am so proud of her strength and commitment. Those first few days, the outlook was grim. It must have been incredibly difficult, but she found the time to coordinate homelife, make sure the kids were taken care of, and keep friends and family updated about my condition. And somehow found CaringBridge.com, set up and account, and started chronicling everything.

3. May

Thursday, May 26, 2011 12:16 PM, CDT
Sorry to keep you hanging!

Took awhile to get back to a computer.

Jay's potassium levels were better at 2am, so they started warming him up. As of 6-6:30 this morning they had him back to normal and took that machine out of his room.

I was typing the backstory while Ben had his physical therapy appointment (whole other story) and had just finished the backstory when it was time to go. Dropped Ben off at school and got to the hospital, where Judy gave us very good news.

They had turned off the paralytic med and waken him up for an assessment - he was able to open his eyes, respond yes and no with his head, indicated he wasn't in pain, squeezed the nurse's hand when asked. All wonderful signs!

When I got here he was back to sleep, that's still better for him right now. I don't think the doctors have been back (if they did, they didn't stop by the waiting room) to figure out when to start the next steps. The nurse thought the next step would be to remove the balloon pump in his aorta which has been helping support his blood pressure since he was at Methodist. There's been some bleeding from where that tube goes in, so that would be a good thing to get rid of.

It is appropriately a sunny day in Minnesota which goes with the continued good news. There's still a long way to go, but the steps are all going in the right direction so far.

> I only have the vaguest recollection of someone trying to get my attention, calling out *"Jay? Jay, can you hear me? Can you squeeze my hand?"* I was in a room filled with beeping noises, faceless people, strange machinery, and countless tubes. Everything looked grainy and black and white, like in a classic

silent film. Something in my mind kept telling me it was just a dream and to go back to sleep.

And so I did. Occasionally, I'd be brought out of sedation to see how responsive I was. Invariably, some complication would arise and the doctors would put me back under again. I don't recall much of anything I know for 100% certain is "real" from here for these first few weeks. But my mind started to weave together a series of bizarre, outlandish, and often horrifying realities. I'm guessing it was all in an attempt to somehow rationalize the various stimuli I was bombarded with, and the snippets of conversations or noises I would hear around me.

This is where things started to get weird for me. Yeah, like they were totally normal up until now.

Thursday, May 26, 2011 3:53 PM, CDT
Wow

Met with the doctors this afternoon, listening to them basically recap what Jay's gone through was a little overwhelming. I'm glad I didn't realize quite how close things had gotten until they weren't quite as bad.

Things are still going about as well as can be expected, it seems they've been varying his sedation level because he was moving his legs, he again opened his [eyes] yet again when I talked to him.

Nothing dramatic should happen over the next few days. The long-range plan will be to try to reduce the extra support he's getting and try to determine how his body reacts.

They aren't (quite) as worried about his heart - if that's been weakened, they have options to help it. They are still pretty guarded about the length of time his brain wasn't getting enough oxygen. The signs so far have been good but they can't really do much until they can take him out of sedation, and they can't really do that while he's hooked up to everything - if he moves wrong, bad things will happen, so naturally they're trying to keep him from moving much.

Honestly I think they're surprised he's responded as he has.

That's it for now. Still gonna be several more days of sitting around. At least we remembered to bring cards today.

> It's difficult for me to sync up my flurry of memories, alternate realities and waking dreams with what was actually going on. In addition to suffering memory loss, I also lost almost all sense of time and sequence for quite a while. Since I don't know what order these delusions came in or all the details, I'll describe several of them in a separate chapter.

Friday, May 27, 2011 9:37 AM, CDT
Not a good way to wake up

...so it's a good thing I wasn't sleeping when the phone rang at 5:15am.

Jay's potassium levels had continued to climb overnight, and his kidneys weren't functioning well enough to process it all out, so the doctor needed permission to start hemodialysis to help clean it out.

Now that rush hour is over we should be heading over to the hospital soon. Will try to post another update if there's any new news.

Thanks to everyone for their offers of help - at this point I don't think there's anything I need that my current support (my parents, Jay's mom and brother, and Jay's dad should be here on Sunday) can't help with. But I will certainly keep you posted. Thanks for all the prayers too.

Friday, May 27, 2011 5:19 PM, CDT
not all news at 5:15am is bad

OK, so after getting to the hospital around 10:30 and having an in-depth talk with the doctor, the dialysis wasn't as scary as it seemed. It's kind of a "normal" thing - once he was warm and had better circulation going, his cells released a whole lot of trash in the form of potassium and phosphorus. Potassium is good, but not too much. It gets removed by the kidneys. Unfortunately, the timing of stuff is that the kidneys react a little slower than the rest of the body, so they were reacting to the stress from earlier in the week, and they just couldn't keep up. So the dialysis was just a little boost, and they were wheeling the machine out of his room as we arrived.

Other than that, things are going about the same. His kidneys are still a little bit under par, but they said that's really to be expected, and should hopefully recover in a day or two.

Once they do, then the next step will be a "turndown TEE" - they will start turning down the support provided by the ECMO and see how his heart reacts. If it does ok with the first small step, then they will do an ultrasound of his heart via his esophagus. That will let them see how his heart responds to the extra work it needs to do. And once they know that, they'll have a much better idea of what comes next.

One of the options will be an "LVAD" - basically an implanted pump that will do the work his left ventricle might not be able to do. We met with the nurse coordinator for that program and he was very informative and forthcoming in answering all our questions. It really could be up to a week before they even make a decision about this, so still lots of wait and see and hope.

That's about it. Thanks to the FFG crew for the card, I read him some of the comments even though he's still sleeping.

> My friends and co-workers from Fantasy Flight Games were awesome. I received cards, encouraging phone calls, and personal visits from many of them. In addition to Trish and our immediate family, I am blessed to have a strong support network we can rely on.

And now to try to relax tonight with pizza and a movie at home. This is going to be a loooooooooooong journey.

Saturday, May 28, 2011 3:21 PM, CDT
baby steps

Still making baby steps today. Otherwise not a lot else to update.

They've turned off his sedation to see if they can get him to wake up, but as he's been down pretty far for pretty long they're not surprised it's taking a while to wear off.

Hopefully he'll come around enough before Chris needs to go back to the airport.

Still in wait and see mode, not likely to be much happening over the weekend. Just hoping for more and more baby steps continuing in the right direction.

Everybody get out and enjoy the holiday, thank a veteran, and hug your loved ones.

Sunday, May 29, 2011 8:26 AM, CDT
Sunday Sunday SUNDAY

(you have to hear that in the monster truck guy's voice)

Still not much new - I called his nurse for a morning report. They're still keeping the sedation off, he's opened his eyes and moved, but not yet in response to commands.

Marilyn (the nurse) said that's really not unexpected, given that his kidneys aren't working at top efficiency it'll take longer to clear the medication from his system.

About time to head to church and do some extra praying, taking the day as it comes without any set plans (sort of getting used to that, but I'd really like to be able to get back in a routine).

Sunday, May 29, 2011 7:50 PM, CDT
Sunday evening

Visited again today. He's still off sedation but still sleeping. He did open his eyes once when asked.

The ECMO support is down to 2.5 liters/minute and he's still doing fine. Kidney levels that have been worrying are still high, but have leveled off over the last 12 hours; hoping that means they'll start coming back down soon.

Tomorrow morning they will do the echo study - turn the ECMO support way down, and watch what his heart does via ultrasound. That will give them an idea of how strong it is and how much extra help it might need going forward.

Then on Tuesday they should have enough info to decide next steps; best case being that his heart is strong enough that they can take out the ECMO and he doesn't need an assist.

The LVAD is still a possibility - they'll know a lot more once they do the study tomorrow.

Baby steps continue in the right direction, but still a lot of woods to get through.

Monday, May 30, 2011 3:13 PM, CDT
Happy Memorial Day!

Had another good visit today. They were able to do the echo this morning and everything went well. Depending on the availability of the surgeon and an OR tomorrow they should be able to remove the ECMO machine; if not tomorrow then Wednesday. They'll be able to do a more extensive echo in the OR for a longer time, so they'll be able to tell if his heart is ok on it's own (though damaged) or if he'll need the LVAD for extra support. That's still TBD.

Finally had the brilliant idea to bring a book tomorrow that I can read to him. The waiting room is getting pretty uncomfortable and not very exciting. :)

> That was really sweet of her. I wish I remembered this, or knew if anything she read to me influenced any of my subsequent delusions. I'm sure on some level I heard her and took comfort.

Tuesday, May 31, 2011 11:29 AM, CDT
wait...and wait...and wait....and GO

Not sure where I left off last night - the plan was to unhook the ECMO today in the OR. They didn't know when yesterday - they were waiting for the surgeon to get out of surgery. I called for an update last night - he was STILL in surgery so they still didn't know if it would happen Tuesday or Wednesday.

This morning's update was that he had started moving around, pretty agitated so they had to sedate him again. They brought in the dialysis machine for a treatment (there was debate yesterday between the kidney doc and the heart doc over whether it was needed, I guess the kidney doc won). He was "penciled in" for the OR after the current case (this was about 9:30am) but the nurse didn't think he'd make it down until after 3pm.

After taking care of some stuff at home I rolled in here around 11:20, got set to turn my phone off to hang out in Jay's room with him. And noticed the text from Chris - he'd called the room and found out they were about to take him down to the OR. Yes indeed, he was gone from his room, they were ready to go, and "it shouldn't take long". In hospital terms, I'm not sure what that means but I'm hoping it's no longer than 3 hours.

That's all I know right now.

Tuesday, May 31, 2011 12:44 PM, CDT
back in his room

Jay's back in his room, the ECMO is gone. And the bleeding he had from the "entry sites" is done also.

He's still a little jaundiced - his liver 'numbers' are ok, just has high bilirubin, partly from the blood products he needed due to the bleeding, so that should start to clear up now that he's not bleeding and won't need any more blood.

No LVAD yet, the surgeons said it might even be a month or so before they decide if he needs one (I'll see what the cardiologist says later today, I'm sure). The front of his heart isn't moving very much but it seems the rest is doing well enough that he doesn't need any extra help yet.

They may take the balloon pump out tomorrow, and they said something about working on getting him off the ventilator as well, but obviously are taking things one step at a time, so the rest of today is recovery and rest from surgery, and we'll take tomorrow as it comes.

The surgeons said to expect him to be in the ICU for at least another week, but in the meantime I'm just grateful for the minor and major miracles God has sent our way so far.

And for all the love and support from near and far, family and friends; I couldn't have made it through this week without it.

> It is frightening even now to think of what that first week must have been like for all of the people witnessing it firsthand – the doctors, nurses, family members and especially Trish.
>
> It is also hard to imagine that I was completely reliant on so many different machines to fulfill basic bodily functions. Once I received my medical records, the full impact of how dire the situation was really hit me. The doctors did an amazing job holding me together long enough for my body to start recovering.

4. June

Wednesday, June 1, 2011 8:43 AM, CDT
Wednesday morning update

Yesterday afternoon they did an abdominal ultrasound to check out his liver, etc. and it came back normal.

He did also respond to Dr. Martin when she told him to squeeze her hand and blink to let her know he was listening - he hadn't done that in awhile so it was very good to hear that. Then they put him back on sedation to sleep for the night.

> Ah yes, another moment I don't quite remember. At least, not at the time. However, a few weeks later, I was lucid during one of Dr. Martin's visits, and when I saw her it was like being hit over the head with a sledgehammer made of déjà vu. A flood of images washed over me. It was *her* face I had been seeing in my dreams and fuzzy memories, standing over me, asking me if I could squeeze her hand or blink my eyes. What a surreal experience to "see her again" for the very first time.

He's still on sedation this morning, sounds like they will keep him asleep until they remove the balloon pump which should come out sometime today. Then they can shut off the heparin (blood thinner), too. He'll be laying flat for at least 6 hours after that, and then they may also remove the breathing tube.

So it's entirely possible that by the end of the day he'll be on IVs only - I won't say likely...one thing at a time. But compared to a week ago, just the possibility is wonderful.

The only "but" is that his kidneys are technically in failure, but that's an acute condition caused by the heart attack and shock. The doctors and nurses still have every reason to believe the kidneys will recover and be fine, but he may still need some occasional dialysis until they do. His nurse said she hadn't heard any talk about that yet today, but they did dialyze (yes, google says that's a real verb) him yesterday before removing the ECMO.

It's still going to be a long journey out of the woods, but he's on the right path and moving slowly in the right direction. And the trees might be starting to let a little bit of light through.

Wednesday, June 1, 2011 3:29 PM, CDT
Wednesday afternoon update

Balloon pump is gone!!

He's aware enough to know that he really doesn't like whatever's in his throat (the breathing tube) so he was getting pretty agitated over that. Which is, ironically, a good sign, except that they'd really like him to hold still a little more until the site the balloon came out gets to seal itself up a little more.

Sounds like they are going to wait and get the breathing tube out tomorrow morning; although as always that's subject to change without notice.

That's it for now. I'll probably make one more check back in the room before I head out from the hospital and pick up the boys for the night.

Thursday, June 2, 2011 9:44 AM, CDT
Slow day

They brought his sedation off a bit so he could wake up - he was wiggling toes and squeezing fingers again. They did a "pressure support" test on the ventilator and his breathing got too fast, so the tube stays in for now. They'll try again tomorrow.

He's also a little "up on fluids" (his ankles have been looking a little puffy) so they're giving him lasix to get that out, and they'll probably be doing dialysis again today.

It's cloudy today, which isn't helping me shake off being really tired. Nap before I visit, or visit first and come back to nap before I pick up the boys? Maybe I'll let the magic quarter decide.

Thursday, June 2, 2011 6:28 PM, CDT
good day anyway

So it turned out a pretty good day after all, even with the breathing tube still in.

They decided he didn't need dialysis today (maybe tomorrow, we'll see). He gets pretty restless and antsy as he gets more awake, so they've got him back on enough sedation to keep him calm. The nurse dialed it down a bit when I was there - she wanted to see if he'd respond to me and he did squeeze my hand! Yay!

He was going to get a PICC line when I left (direct line through his arm to almost his heart, for meds if needed), which would let them remove the swan line from his neck serving the same purpose, but in a much less comfortable fashion.

> The scars from the PICC line and SWAN GANZ line are still visible, but only if you know where to look. They are just two of the scars from the ordeal that are still visible. It never ceases to amaze me how few physical reminders there are of everything I went through.
>
> I am grateful for that, though. The scars I have are plenty.

Improvements to liver and kidneys are small but steady. I'll take it.

They are going to try to get the breathing tube out again tomorrow. The "pressure support" basically sounds like they turn it off (after waking him up a bit) and see how he responds - a familiar process. But this morning he started breathing a little too fast for comfort, so they decided to wait. It could be that he's "one of those patients that gets anxious when it's turned off", in which case they might take it out anyway to see if he calms down then. (Jay? Anxious?? Never.)

I was almost able to say that with a straight face. Almost.

Friday, June 3, 2011 9:25 AM, CDT
16 years

Happy Anniversary Jay!

This year counts as "sickness" and "worse", so next year better bring "health" and "better"! "Richer" wouldn't hurt either. :)

> I was absolutely heartbroken when I found out I had missed our anniversary while unconscious. But Trish cheered me up by reminding me I was lucky I didn't *permanently* miss it, and was still around to celebrate. And of course, I would have to make up for it next year. Thankfully, I had planned ahead and gotten her gift early. It was already stashed at home. Once I was thinking more clearly, I was able to remember the gift and tell her where to find it. She was pleasantly – if a bit belatedly – surprised.

Saturday, June 4, 2011 9:21 AM, CDT
Saturday the 4th

Had a headache when I got home yesterday, sorry no post. Took a nap, Ben took me out for anniversary dinner. Nice evening.

Today's status is still much the same; they are keeping him mostly sedated to rest, and then about 4 times a day they are going to pressure support - turning off the vent and letting him breathe on his own. He was doing well on it this morning when I called, and seemed less agitated on the lower sedation than he was yesterday according to Jack.

Jack and Judy are beginning their trek to WI today to visit family, Steve and Shelley came up last night from IA to hang out for the weekend, then Joe and Mary should arrive Sunday for their support shift.

So this might be the week for patience if he doesn't get off the vent this weekend. And if he doesn't, they were talking maybe a trach tube on Monday - it's not good to keep the tube down his throat much longer as it can start to cause problems with his vocal cords. But that's not such a bad thing, it'll be a little more comfortable for him if he needs the extra time for recovery.

Guess that's it for now...patience is a virtue.

Saturday, June 4, 2011 7:37 PM, CDT
still making steady progress

They've given him two tests of turning down the vent to see what he can do on his own, and he's done great both times today. They're probably in the midst of a third, based on their plans when I left. If he does good in that one, and in the one tomorrow morning, they might just get the breathing tube out tomorrow. Fingers crossed - then they don't have to keep him sedated (he really really *really* hates the tube when he starts waking up).

They also moved him into a chair for a bit this afternoon, sitting more upright than he can get in bed is also good for his lungs.

> Since I could not walk – or even stand – on my own yet, they had to use a special device called a Hoyer lift sling to hoist me out of the hospital bed and swing me over to the chair, where they would carefully lower me to sit down.
>
> Here are two things you probably didn't know about the Hoyer sling system. One, the sling uses hydraulics to lift you. Two, if the straps on the sling aren't properly secured, genitals will be crushed when the sling jerks upward and your entire body weight is dropped on them.

His creatinine levels are still pretty high (kidneys aren't working real great) but they did come down a "tisch" which sounds like a medical term for skosh. The kidney doctors were ok with the way things were going and didn't think dialysis was necessary.

He's looking much less like a Simpsons character, so I have to assume that means the bilirubin levels are headed back to normal.

Here's hoping maybe I can see him tomorrow without the breathing tube. I know I've been saying that for a few days, but I'm still hoping!

Sunday, June 5, 2011 8:55 PM, CDT
what a day

I called for the morning update today and he was still not doing his breathing tests (aka pressure support, aka weaning trials) very well when they lifted the sedation a bit.

He was fine when on the sedation, but that wasn't a very good test for them. But they planned to keep doing a few more during the day, and getting him back in the chair a few times, but didn't expect much progress.

Between one thing and the other, and taking a nap when Steve and Shelley took the boys to Nickelodeon Universe, I didn't actually make it to the hospital today.

> We are eternally grateful to our families for dropping everything and helping out for as long as they did. Especially by keeping the kids distracted, so Trish could focus on work and me.

But I did get a phone call from Dr. Martin later in the afternoon (around 5pm) - she's been his attending physician and has been great at keeping us informed of both the good and bad news. She said she had a surprise for me, hold on just a minute.

She put Jay on the phone, and he said hello! He said a bunch of other stuff too, but he was still pretty sleepy and mumbly, and there was static on the phone so I couldn't really tell most of what he said. But it was still wonderful to hear his voice!

He had been continuing to do well on the breathing trials so they decided to pull the breathing tube out, and even though he was still pretty tired he was doing just fine.

> I do not know if I am remembering this properly or not, but I recall a red-headed nurse standing at the end of the bed, whose sole purpose seemed to be getting me to focus on her and not what the doctors and other nurses were doing as they removed the breathing tube.

> I felt a slight tug, then an intense gagging sensation. It felt like I was being forced to vomit one long, continuous exhalation as they kept pulling the tube out of my throat. I know the tube wasn't very long, but at the time it felt like a magic act, when the magician pulls the handkerchief out of his pocket and it just keeps coming and coming.

So I still haven't seen him but I'll be heading over first thing in the morning to go talk to him!!

Monday, June 6, 2011 2:57 PM, CDT
Monday Monday

Things are still going pretty well. He's more awake today, but tired, and his voice is barely a whisper over the oxygen mask.

But they pulled out two more IV lines - one in his neck that wasn't needed anymore, and one in his hand that was monitoring his arterial blood pressure (don't ask me how).

> I had IV rigs in both arms, and felt a bit like a marionette. But they still had to draw blood for more lab tests, and soon they had to start looking elsewhere. Whenever they'd come in for a blood draw, they started poking around on the backs of my hands and feet. The more weight I lost, the deeper they had to dig to draw blood. Ouchies.

A Speech-Language Pathologist stopped by to do a swallowing eval. Jay was very happy to have some applesauce. He's now eagerly awaiting her to finish her documentation and recommendations so he can maybe have some more food today.

> I remember this well. I had been on a feeding tube since arriving in the ICU, and had nothing to "eat" but Ensure pumped directly into my stomach. That first taste of applesauce was... divine. It was the sweetest, most amazing thing I had ever tasted. It tasted so good, I actually cried.

His sodium is still high, they are treating that. His kidneys are still slowly recovering. He's still pretty weak-kitten and tires easily, not surprising after 2 weeks in bed.

But he's in good spirits, still a little groggy. I haven't heard any new plans, so I'm going with the continuing plan of baby steps, don't push too hard, and let him recover some more.

Tuesday, June 7, 2011 5:43 PM, CDT
Two weeks

Two weeks since Jay got transferred to U of MN. The progress has pretty much been remarkable.

Updates are probably going to be slower in coming now, most of the big hurdles are done.

His kidneys are continuing to get better, the creatinine levels are dropping. His sodium is still a little high but apparently that was at least partially due to the tube feedings they had him on, they changed to a different kind that should be better for him.

Respiratory is still a little weak, so they're giving him some nebulizer treatments to open up his lungs and break up the "secretions". Not unexpected, just takes time to build that strength back up so he can take the slow deep breaths they want him to.

> It was so hard to breathe once the nebulizer treatment started to wear off. At the time, I only felt like I could inhale and take full, complete breaths when the nebulizer mask was fitted over my face, and the moist air soothed my throat.

PT and OT were in again today to help with the physical strength. That's going to take a while to rebuild as well, but he seems pretty determined to work hard even though it wears him out pretty fast.

I missed the doctors today so I'm not sure what long-term plans are, or even short-term plans beyond continuing the status quo since it is working. It's just going to take patience, so hopefully that's something Jay can finally learn!

Wednesday, June 8, 2011 8:46 AM, CDT
Late night phone call

No one ever calls at 3am with good news.

Jay's breathing was getting more rapid through the night, and his chest x-ray shows a lot of fluid, so they've got him on more antibiotics and back on the vent and breathing tube. They've got him sedated again also.

They are currently putting a catheter back in his neck (swan-ganz for you healthcare folks out there) that can let them measure the pressures in his lungs and heart. This will tell them whether the fluid is caused by an infection in the lungs or if the heart isn't strong enough to clear the fluid out. Then we'll have a clearer course of action.

> I cannot help but think this may be the reason that suffocation and drowning were such common themes in my delusions.

Pulmonary will also be in today to get a sample of what's in his lungs to get more info about what's going on.

More to come as they figure it out and let us know.

Wednesday, June 8, 2011 12:56 PM, CDT
sitting and waiting

They had problems inserting the swan-ganz line in the room due to scar tissue, so they took Jay down to the Cath Lab about 10am to put it in there. Someone else showed up that was sicker, so they just got started around 12:30 with Jay.

They'll also be doing a chest CT and a bronchoscopy to get better looks at his lungs to see what's up. Expecting him back in his room about 2ish?? But that's hospital time, so who knows?

Also discovering there's only so many games of solitaire one can play. Or two, trading back and forth.

Wednesday, June 8, 2011 5:36 PM, CDT
good, bad, yet to be determined

Bad - The CT showed a lot of fluid in his lungs.

Good - The pressure study showed that it's not because of his heart.

Good - The bronchoscopy showed that it's not blood in his lungs.

Yet to be determined - Exactly what the problem is. A sputum culture from a few days ago came back today showing positive for candida - a yeast infection. They've got him on specific antibiotics for that, as well as some others to try to hit likely and/or most potential targets.

> The fluid build-up in my lungs was disheartening. After so much progress, this felt like such a huge step back. Or it would have felt that way, if I had any idea it was going on.

So they'll culture the sample they took today of the lung fluid, and they'll culture the blood samples they took today.

Next steps were to put in an arterial line again, this will let them measure blood gases. Then they'll try to wean down the vent a bit. And wait for the results.

If it's not just an infection, it could be respiratory distress syndrome, which, quite frankly, is something I wished I hadn't googled. It didn't sound so serious when the nurse practitioner mentioned it was a possibility.

Never thought I'd be hoping for a yeast infection.

Thursday, June 9, 2011 4:01 PM, CDT
no news today

No expected news today. Just waiting for antibiotics to work, and cultures to grow. They are "supporting him through this."

So no news is good news.

Friday, June 10, 2011 12:25 PM, CDT
another slow news day

Small improvements, his lungs are clearer and they've taken him off the nitrous oxide that was helping him "oxygenate". They've tweaked one of the settings on the vent as well.

Most of his "numbers" are getting better, with the exception of hemoglobin. They gave him a unit of blood this morning.

He's still running a bit of a fever. Still sedated but less so, and while he opened his eyes and wiggled his toes a bit he's not really awake enough to follow commands.

> The whole "fever dream" state could also account for some of the bizarre delusions. Fever dreams are crazy enough when you are not in cardiogenic shock.

If today continues well, they might start trying to wean down the vent again tomorrow.

Saturday, June 11, 2011 11:42 AM, CDT
Day 19

Still sedated, still on vent. I think they were waiting on some labs to determine if they could tweak the vent settings anymore, try to do a pressure support test.

He nodded his head yes to the nurse who asked if he was in pain, so they upped the pain med a bit.

> When I would start to wake up, I was almost always in intense pain. Most of it along my right side. My right arm and leg had suffered severe neuropathy, or nerve damage. I lost the use of my right arm for several weeks. Eventually sensation started to come back, but as my arm and leg recovered, it felt like swarms

of angry bees stinging me constantly. There's still a dead zone on my right leg where the ECMO surgeries were performed.

Creatinine and sodium are still high but coming down.

Still nothing to do but wait and see.

Monday, June 13, 2011 6:09 PM, CDT
sorry for the long pause

Sunday's update was no update, and I was kind of in don't-wanna-do-anything mood so I didn't make it downstairs to post.

Today's update was delayed as I've started back to work in the afternoons. I'll hang out with Jay in the morning and talk to the docs as they make their rounds, then come home, change and go to work. At least that's the plan.

It felt surprisingly good to be back at work today - it seems my limit of not doing anything in particular is 3 weeks. And I actually got caught up on most of my emails - read them all anyway, didn't resolve them all.

So on to the update:

The sample that they took last week via bronchoscopy did turn up something finally, that must have been yesterday when I didn't call for the evening update. It grew one colony of VRE - vancomycin-resistant enterococcus. Doesn't sound like a particularly nasty bug, antibiotic resistance aside. They've got Jay on something that knocks it out, and meanwhile they suction his lungs every so often since he can't exactly cough all the associated "secretions" out. Doesn't that sound great.

> There is actually a silver lining to the VRE infection once I got to Regency Hospital. Well, sort of. Everyone who entered the room had to put on a bright yellow gown. Everyone looked like Big Bird, and the boys were just downright adorable in the adult, over-sized gowns.

Otherwise everything else continues well, his heart is holding its own, kidneys and liver are recovering, just gotta clear up the lungs and the extra fluid in his system - a side effect of the high (but getting lower in a very controlled way) sodium. I think his ankles are puffier than mine were at the end of pregnancy, but his hands are less puffy today than they were Saturday.

Tuesday, June 14, 2011 12:49 PM, CDT
three weeks

Yep, it's been 3 weeks.

Jay's lungs are clearing up, both by listening and by xray. That's good news.

They're still trying to get his sodium down, the problem is that if they wash out a lot of fluid to get the sodium out, it washes out potassium too and then that gets low.

They're working on the happy medium.

They put some blue ankle boot-things on him to keep pressure off his heels.

> I do not remember those blue boots, but there were some leg wraps used later. They were... odd. It felt like a warm, deep muscle massage along my feet and calves at all times, but it made my feet incredibly hot. And I swear my feet were moving on their own. I think I told Trish it felt like a herd of kittens were swarming around my legs.

And the heart team is asking an ENT to come by this afternoon and see Jay. The new attending (we'll have a new one every week, unfortunately I didn't see this one enough to remember her name yet) thinks it would be better for him to get the breathing tube out of his mouth and do a tracheostomy. It sounds scary, but it will let him be more mobile, do rehab, less sedated, sit in a chair, etc.

If they decide to do that, he'll still be on the vent, they'll just hook it into the trach tube instead. And then they'll wean it down from there.

And I finally picked up his iPod from his desk yesterday and brought some "real" music to him. The breaking point for me was that they had him listening to Lite FM and someone asked if he liked Elton John.

Not that he has anything against Elton John, he just wouldn't show up in Jay's Top 50 list. Maybe not even the Top 100.

> Lite FM piped in drove me crazy. Bring me my heavy metal! Where is my Overkill, Voivod, Megadeth, Anthrax, Metal Church, Opeth, Lordi, Metallica, Manowar, Candlemass, or Iron Maiden? There is something very cathartic about heavy metal music. Then again, I am just as likely to listen to They Might Be Giants, Delerium, Johnny Cash, Billy Joel, Weird Al Yankovic, Depeche Mode or Chris Isaak. But not Lite FM!

Wednesday, June 15, 2011 6:29 PM, CDT
still taking baby steps in the right direction

and still a long way to go.

Jay's continuing to improve day by day, just hard for us to see since it generally means tweaking settings on the vent.

They'll be putting in the trach tube tomorrow morning; he'll still be on the vent but without the tube in his mouth they won't have to keep him sedated to be comfortable, he'll be able to move around more, maybe get back in the chair instead of in bed all the time, and rehab can come back in and work with him.

Spoke with the doctor today to see what might be a little further down the road than taking one day at a time, these are her best guesses and are subject to change without notice (like everything else).

- Another week to week and a half in the ICU.

- If they get him off the vent, or onto a type of vent that doesn't need as much support they can move him up a level (or down a level...inside joke) and move him to either a step-down unit or just a "regular" wing of the hospital - get him out of the ICU.

- He could be in the hospital for another 3 weeks.

- He'll probably need to go to inpatient rehab after that, and then of course outpatient rehab for some amount of time.

- Could be another month or month and a half before he gets to come home.

Yikes.

> During one brief period after I was brought out of sedation, someone asked me *"Do you know what season it is?"* My response was immediate. *"Baseball season."*
>
> Apparently that was not the answer they were looking for. This was sometime still early in the 2011 MLB season. When someone told me the Cleveland Indians and Kansas City Royals were leading the AL Central and the Twins were in last place in the division, I figured I must still be in some strange, upside-down world and settled back into sweet oblivion.

Thursday, June 16, 2011 7:53 PM, CDT
trach is in

They put in the trach today, it was nice to see Jay's face again. (The mouth breathing tube was held on by lots of white straps.) He was still pretty sleepy, recovering from the surgery. He also coughed a lot, which didn't look very comfortable for him whatsoever. But it was clearing stuff out, so there was some good in it.

> Coughing. Oh, the coughing. I had coughing fits constantly. It is surprising how exhausting coughing is when you are already weak. And coughing with a tube sticking out of your throat is painful, too.
>
> When I could not physically cough up whatever was aggravating me, someone from respiratory would come in and slip off my trach apparatus long enough to put a little suction tube into the tracheotomy stoma to suction it out. It was weird to think that they were essentially vacuuming my throat.

Talked to his nurse a little bit ago, he's continuing to improve. He's coughing much less, he was up in the chair again for awhile. The trach site was starting to bleed a bit (imagine that, on aspirin and plavix) but they put this funky dressing on it that makes it clot. I didn't fully understand. The nurse thought it was cool, so it's cool.

And that's all for now.

Friday, June 17, 2011 7:07 PM, CDT
Friday the 17th

That's not supposed to look ominous, I just couldn't think of any better title.

Spent the morning at the hospital, but didn't get to see Jay much. Spoke with nurse care coordinator about the potential to transfer Jay to a long-term acute care hospital, and what that means, and why.

Basically, once Jay's "medically stable" but still on the vent, he doesn't warrant being in the ICU anymore, yet he doesn't "qualify" for an inpatient rehab unit because of the vent. Long story short, the LTACH is the one place that can wean him off the vent and start rehab.

Then we met with reps from the two LTACHs in the Cities, which also happen to be the only two LTACHs in the state. Got a much better idea of what they can do for Jay, and a much better understanding of how this is a step forward in his treatment. (I've used this before, Order of the Stick fans, but he's going up a level by going down a level.)

> I love it when Trish talks nerd. It is like we were made for each other. In some secret government lab or something.

That still won't happen until the doctors here say he's ready to go, but that could come early next week and maybe as early as Monday.

Meanwhile, they've been getting him out of bed and into the chair, they've started some pressure support trials. And they had a "trach dome" (if I heard that right) in the hall that they were waiting for Respiratory to hook up - it would provide oxygen, warmed and humidified, thru the trach, but not pressure as the vent does, all the breathing would come from him. That would be part of the weaning process, an hour or two on his own, then back on the vent when he got too tired.

And that, dear readers, is all I know. Will be taking pics of him and the room tomorrow to prep the boys for visiting so they know what to expect. Hoping for a nice Father's Day visit on Sunday.

Sunday, June 19, 2011 5:58 PM, CDT
Happy Father's Day!

The boys got to see Jay for Father's Day! The nurse had given him a shave so he didn't look quite so much like a lumberjack. And he was sitting up in the chair on the trach dome (aka getting oxygen but breathing on his own).

He looked really happy to see them. They brought him some stuffed animals to keep him company - Nate gave him Octi the Octopus and Ben gave him his very own Teddy. Then they got distracted by the cartoon on TV and couldn't be pulled away.

> I cried like a baby when they came to visit. This was especially moving because when I first regained consciousness, there were a number of photos taped above my hospital bed... and I had no idea who they were. I felt like the worst person in the world when I later found out I did not even recognize my own kids the first time I saw them.

He's been getting frustrated when he can't get a message across since he still can't speak, so I brought in a laptop for him to type on. That didn't work so well either, with his weak arms and hands he had trouble getting the right keys. You'd think hospitals could get special super-big keyboards to help patients communicate better.

He started to get pretty wiped out, so the nurses put him back in bed and we said goodbye. Hope he got some good rest. You can check the Photos section to see if I figured out how to post the pics Jack took. No promises you'll find any, I haven't tried yet.

Tuesday, June 21, 2011 12:10 PM, CDT
Lots to catch up on

So last night at around 7ish they transferred Jay from U of MN hospital to Regency Hospital, a Long Term Acute Care Hospital (aka LTAC or LTACH). He's really doing well.

> I vividly remember taking the ambulance ride to Regency. It was warm and muggy outside. The sky was such a deep, intense blue. And the air smelled so... Fresh. After weeks in the sterile, medicinal hospital, this was my first breath of fresh air since the heart attack. I savored it. I think I asked the EMTs if they would wait a minute before putting me into the ambulance just so I could take a few more breaths.

I don't think he's used the vent since yesterday, just using the trach dome for extra oxygen. No continuous IVs but he still has lines in so they can give him injections that way.

Still has the feeding tube in his nose, he's eagerly looking forward to a speech pathologist doing a swallowing eval to see when he can start having real food. He actually wrote a wish list of things he wants - cranberry juice, orange juice, applesauce, ice pops, sherbet. And lemonade with his family. (Awww, isn't that sweet?)

> Aaah. Writing. A cruel invention if ever there was. Writing was one of the most difficult tasks I forced myself to perform. My right arm was still numb and my thumb and first two fingers dead weight. When writing, I could see the pen in my hand, but I literally could not feel it. Moving my hand to write was agonizing, and my arm would tingle and burn for hours afterward.
>
> But I forced myself to write as much as I could. It gave my mind something else to focus on, and I swore to myself I would get my arm to work again. I wrote lists of food I couldn't wait to taste. I wrote down the names of every nurse or technician who came to visit. I wrote thank you cards to everyone who had sent us a card. I wrote notes on what I did each day.
>
> It is fascinating to go back and look at how much I wrote, and how dramatically my hand writing improved in just a few weeks.

> At first, all I could manage were large, irregular swoops and lines, and was frustrated no one could read or understand.

His sense of humor is coming back. The nurse was asking him some questions to check his mental status, and one of them was "Can you hit a nail with a hammer?" He pointed at himself, then shook his head no.

He's realized that he's got some random bits and pieces of memory loss, but it sounds like he filled some of them in with more interesting stuff than was originally there. I had to break it to him that we did not, in fact, own any part of the Minnesota Twins, or the Breyers ice cream company.

> What a letdown. For weeks I had thought we were co-owners of the Twins. And I simply could not wait to be released and return to work at Breyers and get some fresh strawberry ice cream. We had big marketing plans in the works to steal market share from Edy's and Blue Bunny.
>
> And those were the sane delusions.

Thanks to everyone for all the prayers and good wishes! It's been a long ride these past 4 weeks (yes, really) but things are looking pretty great right now.

Thursday, June 23, 2011 6:39 PM, CDT
3C - Care Coordination Conference

Basically everyone got together this morning to tell us their plan for Jay. PT and OT - get stronger, get him doing stuff for himself (wash face, shave, brush teeth, etc.)

The biggest "news" was why the trach is still in, so it was good to hear, even if the answer is that it won't be out as soon as Jay'd like it (which is yesterday).

So when they first put a trach in, they use a tube big enough to basically fill up the windpipe. And it takes 10-14 days for the site to heal up before they want to do anything with it. After that time, it's healed enough to change it for a smaller one, which would let air through his nose, etc.

Once they have the smaller one in, then they can work on getting the speech valve in, which would let him talk again. And then once the speech valve is in, they can work on building up his swallowing muscles so he can eat and drink again. Then they "cap it off" completely to see how he breathes on his own for a bit (day or two) and if that all goes well, then they can remove it completely.

> The whole business with the trach, respirator, and breathing tubes was stressful. I would go from breathing on my own to having tubes forced down my throat, then having a stoma cut in and a tube forced *through* my throat – it was discouraging and frustrating. Would I be able to talk today? Breathe on my own? Roll over without dragging countless tubes along? Wake up choking as my gag reflex fights against the tubes?

So it's gonna be awhile yet, they won't check if they can downsize until early next week, and it'll depend on how much he's healed if they're able to get started early next week, or late next week.

He did "graduate" today from the Special Care Unit to another wing - he doesn't need the constant close monitoring of the SCU so he can move out to a different room.

> This sort of "graduation" was both a blessing and a curse. While it was a relief for Trish and others that I was doing well enough to warrant less monitoring, it was terrifying for me. Each time they removed a monitor or moved me to a wing with fewer cameras or nursing rotations, my anxiety grew and grew.
>
> As much as I disliked all the machines and monitors, they had become my safety net – if something went wrong, I knew the doctors would find out right away. Suddenly, without some of those monitors, I was frightened that I would have another heart attack or complication and no one would know.

Guess that's it. Next step from here might be to a rehab facility, but it all kind of depends on how things go.

Saturday, June 25, 2011 7:26 AM, CDT
One Month Later

This is yesterday's update, I just didn't type it yesterday. While I was visiting Jay yesterday morning, a nurse practitioner from the lung team came in, and after introducing herself said, "I'd like to see if we can switch your trach today."

Jay immediately grabbed her shoulder - she was afraid he was nervous about it. I told her, "You just became his best friend."

> And cue the tearworks again. I was so emotional during my hospitalization. Even just the thought of having my trach out was elating. If I had not been hooked up to so many wires, I would have hugged her.

She had a concern about part of his site and how it was healing, so she called in the wound care team to consult. They arrived as I was leaving, but Jay asked me to come back that night. I didn't want to get my hopes up about them being able to switch the tube out, so I went back after dinner expecting that there was some reason they couldn't do it, and we'd have to wait until early next week as originally planned.

And then I found out he could talk!! They had switched the tube, and when the speech therapist came in to try to speaking valve, she put it on and told Jay to say something.

"Like what?" he said. And it worked! His voice is still pretty raspy as you can expect, but it sounded sooo good to hear it last night.

> Breathing and talking with the trach "cap" over the valve was a weird experience. I felt like a kazoo, as my throat and the cap would hum as I talked. And the sound of your own voice is always weird. Even more so when it is throaty and thready.

They've also approved him to have ice chips, which was the second thing that made his day yesterday. The speech therapist, who will also help him with swallowing, saw his wish list of food/drinks he wanted. She said that was the coolest thing she'd ever seen, and he could pick 2 things from the list to try at his swallowing test on Monday.

> Hey, guess what I did when she told me about the swallow test and my food list? Yep, you guessed it. I cried. The prospect of eating and drinking real food was exciting. Being able to feed yourself is one of those things so easily taken for granted... and when taken from you, is a humbling reminder of how helpless and vulnerable we can be.

Not much planned for the weekend, probably take the boys up to visit today or tomorrow.

OH! And they upgraded his mobility level - they used to use the lift and sling to move him from bed to the chair, but yesterday he was able to "walk" (with lots of assistance) there. He said he still can't stand on his own, so there's going to be a lot more rehab needed, but yesterday was definitely a very very very good day!

> I needed two people and a walker to help me walk, but that was another huge step (pun) toward independence and returning to some semblance of normalcy. The hardest part was actually standing up from the bed – those muscles had not been used for more than a month, and they were *not* happy. Once I was standing, the blood would rush down into my feet and they would burn and sting until I sat down again.

Monday, June 27, 2011 12:28 PM, CDT
O Happy Day

Holly, the NP who changed his trach on Friday, came in this morning and thought he'd be ok to try capping the trach today. Nighttime capping TBD, since he'll need to start using his CPAP again. (Doesn't work with feeding tube, but more on that later.) Liza, his nurse today, came in when the food machine was beeping - the line is clogged. She tried unclogging it with no success, as did another nurse who gave it his best try as well.

Meanwhile, Jill the speech therapist came in to give him his swallow test. Applesauce, water, and graham crackers. The applesauce and water had blue dye added. So he had a little snack, and obviously this was the best applesauce he'd ever had. And practically the only he'd had in over a month (he did get a few bites at the U two weeks ago).

> These three small snacks were like a banquet. I was surprised that after eating just a few bites of these things my stomach ached terribly – it was not used to having to do anything for a month or so.
>
> It was strange to feel textures again, as well as tastes. Outside of these few bites, the only texture I had experienced was the dry roughness of the roof of my mouth... And the only two tastes I experienced were a sickly sweet chalky chocolate from the feeding tube in my stomach, and an odd floral aftertaste I would get each time my IV was flushed with saline.

When he was done, she suctioned his trach (no blue dye) so he has the thumbs up for REAL FOOD! He should even be done with lunch by now. He's starting with level 3, which is pretty much real food, except raw fruits/veggies, toast, etc. Jill would've approved level 4 which is no restrictions, but understandably Jay didn't want to push too fast.

So between being able to eat again, possibly getting the feeding tube removed (they'll see how he does with lunch, but as it is it's pretty useless), and being ready to get the trach capped and go down to just the nose-tube oxygen, he had a pretty wonderful morning.

Not bad considering he left the ICU a week ago tonight.

> Not bad at all. This was one of the biggest days for me. It was hard to not be depressed, worried, and anxious the entire time I was in the ICU. The progress really filled me with optimism for the first time – I was starting to get back to some normal routines, and feeling a bit more self-sufficient. It is hard to explain how big a deal these seemingly simple things are, until you are forced to live without them.
>
> It was probably right around this time that a steady trickle of visitors would stop by. It was great to see people, especially when I would actually recognize them. Friends, family, and co-workers took the time to come visit, and it meant a lot. Not only was it a brief respite from the tedium of lying there, it was nice to reconnect with people and start to feel like I could get back into the real world.

5. July

Friday, July 1, 2011 7:24 PM, CDT
Zoom

That's how fast the week went by!

Update from this morning was pretty great - he's been doing well with his trach capped that they pulled it out this morning! Weirdly enough, it'll just seal back up on its own, no stitches, so for right now he's just got gauze and tape covering up the hole in his neck.

They also took him off the pulse oxygen monitor that clipped on his finger - he'll still get that back at night but he's done with it during the day. They also pulled his PICC line earlier this week, so he's getting to the point where he's not hooked up to anything.

He's still pretty weak and easily dizzy at rehab time, but this morning he was able to sit up at the edge of the bed without being dizzy, and stand up for 60 seconds. That's pretty frustrating for him, but everyone keeps reminding him that baby steps are good, even if he doesn't get there as fast as he wants.

> It was hard to accept sitting up and standing for such a short period of time as "progress." I wanted to be able to do everything again at once. I was angry at my body for acting so stupid – come on legs, you know how to stand! Sheesh, a baby can do it! The delay in response from telling my legs to do something until they would actually move around was aggravating.
>
> But eventually, I was able to turn that aggravation into motivation. I'm nothing if not stubborn. I dug my heels in (literally and figuratively) and would sit there for what seemed like hours, staring at my feet and legs, trying to flex my knee, rotate my ankle back and forth, wiggle my toes, bring my foot up and down, and try to lift my legs off the bed.

The only main "but" is that he's been having pain in his side that he described as a cramp. At first they thought it was just his stomach getting used to working again, but when it didn't get better they thought maybe something with his liver. And it still didn't go away, so they took him to a nearby "regular" hospital for a CT scan this afternoon. They are thinking it could be a kidney stone, but I haven't had a chance to check in with Jay to see if he's done with the scan and if he's heard any results. I'll post when I hear something.

> The short drive to the other hospital for the CT scan was my second chance to breathe in fresh air. The day was much crisper, brighter, and vibrant than during my previous excursion. The hospital grounds and surrounding area have a lot of greenery, so there was that extra hint of life and vibrancy in the air. Almost like that wonderful smell of fresh earth that rolls in right before a big rain storm out in the country.

Friday, July 1, 2011 7:46 PM, CDT
hold that thought

They just picked him up 5 minutes ago to take him to the CT scan...don't expect results until tomorrow.

Saturday, July 2, 2011 6:10 PM, CDT
so that's what's wrong

CT shows some kidney stones. Which aren't fun by any means, but they're small and not blocking anything, so "not serious".

> Yeah, easy to say not serious. But there were a half dozen or so. And each one felt like a little Exacto blade stuck into my side. It was excruciating. It was the most pain I had experienced in weeks – I was so heavily medicated earlier that pain was more of a numb, buzzing background static... This was sharp, acute, and impossible to ignore. Especially when I would be asked to take a deep breath.

Otherwise things continue to go well, his blood pressure is stabilizing (it had been a little low) as he's really focusing on staying hydrated.

Should be a slow weekend, except for Family Lemonade and Movie Time tomorrow.

Meeting with care coordinator on Tuesday to figure out next step from here.

Happy Independence Day everyone! Or at least everyone tuning in from the USA; if you are outside the USA, then just have a nice Monday!

Sunday, July 3, 2011 5:48 PM, CDT
Awesome Day!

Mom, Dad and I took the kids over to see Jay this afternoon. We watched "The Incredibles" and had some lemonade while Jay finished his lunch. Then he got up in his chair-on-wheels (not technically a wheelchair) and we went and sat outside for a while. It was the first time he's been outside in 6 weeks when he wasn't strapped to a gurney for transport.

> This is probably the fondest memory from my entire hospital stay. If not for my hospital gown, it would have been a normal Sunday afternoon with the family. It was a beautiful day outside – there was a slight breeze and it was warm without being uncomfortable and sunny without being glaring.
>
> I was outside, surrounded by my wife and kids, sipping lemonade, and enjoying the weather. Just like a normal family. It may have been my first moment of true peace and contentment since the entire ordeal began.

And before we got there, he had a shower and got to cut his hair (he did some, patient care aide did the rest) and beard and got all cleaned up again. I'll try to get some pictures loaded one of these days, but as I left the camera upstairs and the computer is downstairs, it might not happen for a bit.

Tuesday, July 5, 2011 6:04 PM, CDT
Another great day

Jay (with help from a walker) got out of his room and walked a bit down the hall and back. Maybe 30 feet in all, without getting dizzy. It was a workout for him, no doubt, but the first time he's left a room under his own power since this all happened.

> I was maybe on my feet for five minutes total, but it felt like I had run a marathon. I was so drained after that short walk I had to take a nap.

On Thursday the Care Coordinator will talk to the Rehab folks and see what more progress he's made endurance-wise. If he can handle 3 hours of rehab a day (not in a row) then he can maybe go to an acute rehab facility to seriously work on building his strength back. Bonus is that there's a potential he could move to the hospital I'll be working go-live support at for the rest of the month, so I could spend less time driving and more time seeing him and more time working. But that's probably going to take a continuation of his remarkable gains he's made in this past week - not out of the question.

Otherwise sounds like choices could be a sub-acute rehab where he could go at his own pace rather than needing to meet the 3-hour requirement. Downside is that those are usually also "nursing homes" so he could be the youngest patient by 20-30 years. Or potentially he'll stay where he is for rehab until they get him ready to go home.

More on that Thursday maybe.

Thursday, July 7, 2011 8:14 PM, CDT
Movin' on up

As his trach site is still healing well and his physical strength and endurance is returning, it's time for Jay to move on to a rehab facility to (hopefully) finish his recovery and get ready to come home.

Mentally he's ready to go, medically he's ready to go. The RTs at Regency would like him to stay a bit longer to make sure the trach site keeps healing (they thought there was a good chance it'd be all sealed by now, but not quite), so he's not going anywhere just yet.

> By this time, I was so ready to get out of the hospital and start physical rehab in earnest. But I was still pretty mentally shaken. While the hospital did a great job with my body, treating my mind

was a much, much tougher challenge. I think I was only able to see the hospital psychiatrist three times during my stay, and it was difficult not having someone to talk about the scary stuff with. Trish had been through enough – I did not want to dump all that on her, too.

The plan is that by Monday or Tuesday he can move to the Courage Center, a specialized rehab facility that can help him get strong enough to come home again. It's actually on the way to Regency, so maybe a whole 5 blocks (and 3 very large speed bumps) closer to home.

> Ah, the Courage Center. My experience there was so... unique that it gets its own chapter toward the end of the book.

So with PT and OT taking the weekend off, I'm looking for ways to keep him from being quite so bored with nothing to do this weekend. If you're in the area, feel free to stop by and pay him a visit!

Monday, July 11, 2011 7:03 PM, CDT
waiting and waiting

Waiting for a room to open up at Courage Center so we can move on to the next step. Keep your fingers crossed for Wednesday.

Trach site is all healed on the inside, just a pinhead-sized bit left to heal on the outside.

Jay's feeling well and in better spirits now that the weekend is over - with no rehab to look forward to there's not a lot to do. So he'll keep doing rehab work at Regency twice a day until they can get him over to Courage Center.

7 weeks tomorrow if I'm counting correctly.

Wednesday, July 13, 2011 7:36 AM, CDT
the final countdown

(cue cheesy 80s synth-rock)

If everything goes as planned yesterday, Jay will move to Courage Center at 11am today!

Courage Center Golden Valley
3915 Golden Valley Road
Minneapolis, MN 55422

His typical day will include breakfast, an hour of therapy in the morning, lunch, an hour of therapy in the afternoon, dinner, and an hour of "recreation" in the evening.

He met with the director yesterday and she thinks he may only need their services for a week or two before he can finally come home! Yay! The end is in sight!

Wednesday, July 13, 2011 6:16 PM, CDT
yes he's allowed visitors

Can't find a "reply" in the guestbook, so in response to questions, yes he can have visitors. As of this afternoon he didn't know his schedule yet, but visiting hours are 8am-11pm. He does have his cellphone with him, as well, and calls would be welcome also.

The big change at the new place is that except for being told he needed a 'spotter' to move from chair to bed, etc. until he's cleared by PT, he's pretty much on his own. I think it was a little overwhelming at first - 7 weeks stuck in bed can make a trip to the lounge to read seem pretty odd. I think he felt a little better about it when I told him to think of it as college again - a few classes you need to attend and a whole lot of free time in between.

> The sudden independence was overwhelming. Sure, there were people there to help if I called, but nobody constantly buzzed around like bees. It took a long time to get used to.

It's also kind of an eye-opener that many residents appear to have life-long disabilities (whether they were born that way or happened by some twist of fate), and he's almost independent and ready to be home. Just have to get rid of that almost....

> My first day at the Courage Center was very strange, emotionally. I felt *guilty* at how well off I was compared to the majority of residents. In just that first day as I was introduced to staff and involved in a few small group activities, I met a twenty-something woman who had just recently woken up from a nine month coma after intense brain surgery.
>
> I also met a man my age (also named Jay) with two small daughters about the same age as Ben and Nathan. He had been hit by a train and had his spine and skull shattered. I think his wife, daughters, and mother-in-law were there every day I was at the Courage Center. His mother-in-law and I spent a lot of time talking since our family makeups were so similar.
>
> It was sobering to realize how much more serious and longer-term the impact of some of these injuries had on these residents' friends and family.

Friday, July 15, 2011 8:54 AM, CDT
Alive is Amazing

> From this point on, I start to post quasi-regularly. So my running commentary pretty much ends here; I was able to post what I was thinking and feeling when I needed to most of the time.

** Warning -- This is a VERY long post **

Hello all -- it's Jay! Yesterday was my first day with internet access in nearly two months, and one of the first things I did was come visit this site to see what happened and read all the encouraging comments from friends and family. (The first thing I actually did was check out BoardGameGeek to try and get up to speed, then the FFG website to see what the latest announcements have been).

It was very emotional to read all the kind comments, words of encouragement, and prayers -- from people literally around the world, and a number of grade school and high school friends. All the prayers and well-wishing worked. I know I would not be here now without Divine Intervention stepping in, and God helping direct my recovery.

Believe it or not, this past weekend was actually the first time I began to learn some of the details about what actually happened. I have no recollection of the time between virtually collapsing in the Methodist ER to waking up several weeks later, surrounded by unfamiliar people, and a lot of machines, tubes, and beeping. Reading over these entries and talking with Trish, it seems like I was virtually coma-like and unconscious for the first several weeks.

It took a long time to adjust to my surroundings -- initially I had such a strong startle reflex that any time someone came into vision (especially someone I didn't recognize) I'd start to panic a bit. And whenever I would start to drift to sleep, I'd startle myself awake, afraid that if I fell asleep, I'd never wake up again.

So only now do I know the severity of the situation, and what actually happened during and immediately after "the event." There are even more details -- some which were quite frightening to find out -- that Trish hadn't posted. It is an unbelievable amount to try and process, and quite frightening to think about.

Speaking of Trish. By now even those of you who have not met her personally can tell what an amazing woman she is. She has managed our household, children, career, and still found time to come visit me every single day. And she's handled it all with grace, unbelievable reserves of strength, and devotion. I could not feel more loved and proud of her.

I'd ask that you keep Trish in your prayers as well -- while other people essentially managed my care for me while my body fought to recover, Trish managed everything else. I'm tearing up just typing this because I know there is no way to properly convey how incredible she has been.

I have a number of memories and stories to share. Trish did a great job tracking the medical and care aspects of recovery, while I've been keeping a diary of the more subjective side of things -- what I was experiencing and going through during the various stages of my recovery. Over time, I hope to share some of them with you.

Until then, know that I am recovering well. Yesterday I walked more than 300 feet -- which was amazing, because it was more than twice as far as my previous high score... and it was unassisted! No walker or cane. A huge improvement and a very good sign.

Physically, I lost all feeling in my right arm and hand when I first woke up, and was worried that perhaps I had lost the use of my arm permanently. Over time, and with persistent Occupational Therapy, my arm and hand are back to 90% normal.

Doctors and nurses continue to mention how miraculous my recovery has been. And of course, constantly say "You're too young for this." Trish gave me the perfect response to that (courtesy of the greatest cartoon ever, Phineas & Ferb). Now whenever anyone says "You're too young" I just smile, nod my head and say "Yes. Yes I am."

All this typing has been pretty tiring -- my right hand literally cannot feel the keys, so I have to go back and make a lot of corrections while I go.

Before signing off, I would like to say Thank You -- but that seems so inadequate. The outpouring of love, cheer, prayers, support, offers of assistance, and encouragement is amazing. I am greatly appreciative and humbled by all of this -- I know I am incredibly blessed to be doing so well physically, and that your thoughts and prayers have helped me recover emotionally and spiritually, as well.

Cheers.

Sunday, July 17, 2011 9:21 PM, CDT
Weekend Update

Friday's rehab appt went great, here's Jay's account:

"I walked up and down the hall several times, aced some balance / gait tests -- avoiding objects in the way, continuing to walk straight while looking left or right, stepping over small objects, turn 180 degrees and walk in the opposite direction, etc. And I can tell I'm returning to a more normal, natural, "me" gait -- I'm not locking my legs straight like I was before, and as I get more confident, I'm walking less guardedly.
The best part was the stair test -- the physical therapist said to try and get up the first flight, and she put a chair up there so I could rest. My first

step was tentative... I wasn't sure how it'd feel putting all my weight on one foot while balancing and pushing up. But then, like riding a bike, it just came back to me. I was able to walk up the first flight with ease (not feeling winded, no dizziness), and told her I could do the next flight, as well -- so two flights of stairs (and the mini-landing with three steps) up and down, and probably close to 1,000 feet or more walking... I have been walking a LOT."

Saturday he had another therapy appointment in the morning and walked several laps around the center "square" of the building and several flights of steps.

They're pleased with his progress - not only with how far he's been able to walk and push himself (safely), but his heart rate and blood pressure have been stable.

Not much to report for Sunday, although thank you to Therese (Trish's aunt) for a surprise visit while traveling to see friends in the Twin Cities. Jay really enjoyed the visit and was glad for some company on an otherwise slow Sunday. And hot, and humid. I thought we moved away from St. Louis, but it feels like we're back there again!

We're hoping that the upcoming week's schedule is more consistent, with fewer last minute changes. Hopefully he can focus on what he needs to achieve to come home by the weekend?! That might be wishful thinking...but would be wonderful.

Tuesday, July 19, 2011 9:29 AM, CDT
The Sweet Smell of Success

(The occupational therapist has recommended I take a break from typing, so Trish offered to type this journal update.)

Monday rehab went even better than last week. For my OT session Monday I was put in a kitchen and told to make lunch. And it was pretty much up to me to figure out what I was going to make, where things were, and to get things organized. I settled on pasta primavera, and started to get ingredients together, only to find out that every vegetable in the fridge, except the onions, had gone bad. All the chicken breasts were frozen, so we settled for pasta con onion.

Once I got into the rhythm of things, I forgot about my arms and legs and balance, and just cooked. I love cooking, so this was a great session. Staff walking by would stop in to ask what smelled so good. By the time I was done, I served myself a large helping and sat down to eat and rest.

The OT commented on how well I did - I closed the refrigerator, turned off all the burners, got ingredients out before so I wasn't frantically running around, and I was actually up and on my feet for 30 minutes. I was able to cut the onion using fine motor skills, and the pasta turned out pretty good. I'm glad it did, because lunch that day was terrible, so at least I had something good to eat!

Afterward, she ran me through a series of cognitive and memory tests. I aced them, and feel even more fortunate that the "event" does not seem to have impaired my mental faculties. At the end of the session, she told me that as far as OT is concerned, I'm ready to go home! There are more pieces to the puzzle, but this is a great start.

In the afternoon I had physical therapy, and ended up walking further, climbing more stairs, and performing more tasks than in any previous session. It was pretty exerting, and by the end of it I was exhausted. My blood pressure, pulse and oxygen saturation were consistently good throughout. Sounds like PT is convinced I'm ready to go home as well.

Even when I do get home, there's a lot of work ahead - outpatient therapy, cardiac rehab program, doctor follow-ups and a daily regimen of nearly two dozen pills. Still, that's a small price to pay to be home with my family.

Wednesday, July 20, 2011 8:58 AM, CDT
If it's not one thing...

Jay's right shoulder is even more inflamed and swollen today than yesterday, according to the OT. She's going to consult her colleagues but thinks maybe there's a nerve impingement going on. Possible MRI in the future. Oh joy.

I love you dearly, Jay, but after all this is done, you're not allowed to have any medical condition more serious than a cold for at least 5 years. 10 would be better. :)

Friday, July 22, 2011 12:52 PM, CDT
Good News, Great News, and Several Uncertainties

Last night I finally met with my doctor. After reviewing my (lengthy) medical records for "the event" and conducting a few tests, it sounded like her opinion is that I had both a heart attack (full cardiac arrest) and a stroke at pretty much the same time. She said that my nerve sensations, recovery, complications, and other symptoms are all classic after-effects of a mild stroke. I'm incredibly fortunate I did not suffer any cognitive damage (thanks for all those prayers!).

So the best news of all is after meeting with my doctor, I've been cleared to be discharged! I'll be going home very soon. Hooray!

Already buzzing with excitement, I was very pleased about how my occupational therapy and physical therapy have gone -- both reaching the point that there's nothing specific the Courage Center could provide that an out-patient facility could not.

I had my final OT session today, and was back in the kitchen again -- I made chicken fajitas (well, as close as possible with their limited pantry and supplies) and everyone was stopping by to see what I was making that smelled so good. The best part is everything felt so much more natural -- getting ingredients, utensils, and all the food chopping and cooking didn't exhaust me as it had before. In fact, with a few warmup exercises before cooking, I was on my feet, moving around comfortably for more than 45 minutes, which far exceeded their goal of being able to sustain myself for 30 minutes.

Immediately after that, I had my last Physical Therapy session, where we took a long walk around the facility (it's a beautiful place, lush and green) but it was so humid we headed back inside before we made a complete circuit. We pretty much walked all over the facility, up and down stairs, and re-tested some of the initial evaluation tests that I had failed; this time I passed them all. After back-to-back sessions with a lot of exertion, it took a while to recover, but it's a good sign that I'm getting some endurance back.

There are still a few nagging concerns that need to be addressed -- a number of doctor's appointments, some additional medical tests to see if there are any other risk factors that need to be managed (especially since I've had two heart attacks under the age of 40), lots of prescriptions (14 different meds if I counted right), lots of therapy... So I still have a lot of work ahead.

One of the remaining concerns is the weakness of my right side. Especially my right arm and hand, which are giving me some problems. They've ruled out a rotator cuff tear or a bursitis issue (the cushioning tissue near the socket that helps absorb some of the impact the shoulder joint takes). I had x-rays taken this morning, but haven't heard back yet. The doctor made me a little nervous when she was examining my shoulder and said, "Feel this small little bony lump by your collarbone? That shouldn't be there." She theorized it could possibly a bone spur, a nerve compression, and a number of other medical gobbledygook I could not follow.

The other odd thing about the doctor's visit was the standard reflex test -- where they take that little hammer and see how your ankles, knees, and elbows respond. Left side was fine and dandy. On the right side, nothing moved -- I could barely feel the impact of that hammer, and it didn't trigger any reflexive response.

So I'm praying that my shoulder problem can be addressed with therapy or medication and not surgery -- I don't want to be in a hospital again for a long time!

Saturday, July 23, 2011 10:52 AM, CDT
TWO MONTHS

Two months ago today, Jay left work and wound up at Methodist Hospital ER. Today, he left Courage Center and wound up at home!

Words cannot express.

Monday, July 25, 2011 3:31 PM, CDT
Recovering at Home is Hard Work

I know Trish posted earlier that I was in fact discharged and got to come home on Saturday -- but I didn't think mundane tasks and routines would be so exhausting. I was hoping to post something yesterday, but was completely drained.

Trish also failed to mention that immediately after running over and hugging me when I got home, the first thing the boys said was "Do you want to play a game?" And then started rattling off a bunch of games they wanted to play. In a way, it was like I hadn't been gone for two months.

Small aside about games -- a few weeks ago our area was hit with a series of torrential storms, which flooded the window well in the home office and leaked into the house.

Luckily, most items were saved, but water ended up leaking into (gasp) THE GAME ROOM! The carpeting had to be scrapped, and approximately half of my 700+ game collection is scattered throughout the basement, waiting for everything to dry and for me to get strong enough to fit them back onto the shelves.

I'm walking well, my blood pressure has been fairly consistent, and I've now made all the appointments I've had to set up -- cardiologists, my primary, some additional tests that need to be run, and so on.

The main concerns at this point are:

• Figuring out what's wrong with my right arm. From the shoulder down it still tingles and is numb in some places. My range of motion is limited and my right arm is very weak. They've taken X-rays, but haven't determined if it's nerve-related or joint/muscle-related. Or if it's possibly just a lingering after-effect of the stroke part of my event.

• Seeing how strong my left ventricle is. Sounds like it took the brunt of the heart attack, but is working fairly well "all things considered". From what I understand, that's one of the main things the cardiologists want to look into so it doesn't become a problem down the road.

- Ensuring my vitals remain in the proper range over time. I've got some meds I only take if my blood pressure is a certain level, etc. At the moment, I believe I'm taking between 20-25 pills per day, and check my blood pressure and heart rate every few hours. Trish found a great app for my iPod to track it all, which was very useful when I went to see my primary physician.

I've also been fitted with a LifeVest. It's a portable, personal defibrillator I wear just as a precaution. It's a stretchy fabric harness with sensors and metal pads strung all over to track my vitals. And it automatically sends all the information to the doctors for reference while recharging. So while it's serving a very important function, it's awfully awkward to wear, since several of the metal plates are lined up right along my spine (making it quite uncomfortable to sleep), and I have to carry a "murse" (man-purse for those of you not hip to the lingo) with the battery pack and monitor, weighing a good 3 or 4 pounds.

It's quite the fashion statement. So while at times it gets irritating, I just remind myself that it is a small price to pay for what I get in return -- being at home, sleeping in my own bed, and getting to live and love my family life again!

Thank you all for your continued support, prayers, and well-wishing. I've been told by many of the doctors that my recovery has been truly miraculous -- I'm living proof that all those prayers and positive thoughts work.

6. August and Onward

Thursday, August 18, 2011 7:36 PM, CDT
Time Flies When You're Doing Well

Wow... I hadn't realized it had been so long since we posted an update. That's pretty darn good, since there haven't been any traumatic, bizarre, or unexpected incidents since I've been home. In fact, my recovery continues to go unbelievably well.

I have Cardiac Rehab four times each week, and have been slowly but steadily recovering endurance. Today I was able to walk on a treadmill at a steady pace for 30 minutes without needing a break -- when I started rehab, I could only walk for 3-5 minutes at a time! I'll be continuing rehab for conditioning for a while, and once I reach a certain level of fitness, I'll be able to use our elliptical machine at home and return to a home workout routine.

I'm back to doing a lot of normal, real life things like grocery shopping, laundry, and playing with the kids. And one advantage to the extra time I've had is being able to cook more often. I love cooking, but during a typical work-week, I'd get home around 6:30 - 7:00 at night, which usually meant Trish had to make dinner. It's nice to be able to take time to find heart healthy foods and recipes and feel good about the changes we're making to keep me active and healthy.

I've also been seeing a counselor to help work on the emotional rehab -- the near-death experience and number of complications are tough to put into perspective. And when I was first brought out of sedation and groggy for a few weeks, I had mental gaps, some foggy memories, and other quirks from my ordeal. So it's been very helpful to talk to someone about the bizarre emotional roller coaster I've been on the last few months.

My spiritual rehab is going well, too. I'm grateful that I've felt well enough to start going back to church, and I've made prayer a regular part of my daily life. It's hard to describe how grateful I am for how blessed and fortunate our family has been, despite the grim outlook. I not only survived The Event, but miraculously, I came through relatively unscathed and unscarred.

I didn't suffer any permanent voice damage from the breathing tubes, I've got my mental faculties back (at least, I think I do!), I've recovered feeling in my right leg, and regained use of my right arm -- there's only the barest numbness in the tips of my fingers. Considering all those were concerns along the way, it's amazing to look back and realize I'm still "me" after all this.

I'm still wearing a LifeVest, a portable, external heart monitor and emergency defibrillator. It's a temporary inconvenience. The long-term solution we're looking at is surgery to implant an ICD. An implantable cardioverter-defibrillator (ICD) is a small battery-powered electrical impulse generator which is implanted in patients who are at risk of sudden cardiac death due to ventricular fibrillation and ventricular tachycardia.

I happen to fit that description. Thankfully, I've got a wonderful cardiologist, and we're meeting with the electrophysiologists next week to discuss the ICD and see if surgery still makes the most sense.

Through all of this, though, I have really missed work! I miss being surrounded by all the creativity, energy, and enthusiasm of our workplace. I'm meeting with my boss to discuss getting back to work, and hopefully we'll set up a timetable for my return. Being able to get back to what I love doing will be the icing on the (recovery) cake.

Cheers, and thanks again for all the support, love, prayers, cards, thoughts, and well-wishing!

Monday, August 22, 2011 12:50 PM, CDT
Electrophysiologisticexpialidocious

Even though the sound of it is something quite atrocious!
If you say it loud enough, you'll always sound precocious...

Sorry, but every time I say, hear, or read Electrophysiologist, I can't seem to stop myself from mentally adding the suffix -expialidocious.

We met with my EP this morning and reviewed my medical situation and talked a bit more about the ICD implant. By the end of the appointment, we had already scheduled surgery -- tomorrow, Tuesday August 23!

There was no medical emergency behind the immediacy of the surgery; the surgeon simply had an opening and I want to get this over with as quickly as possible rather than stressing or worrying about it for weeks. Plus, the sooner my surgery is finished, the sooner I can start healing from this, and in the long-term, hopefully fewer interruptions to the rest of my rehab appointments.

As confident as I am with the surgeon and the cardiac care team at the University of Minnesota, I'm a bit anxious about being sedated again -- after all, the last time I was put under a whole bunch of really weird stuff happened...

They're anticipating the surgery to take approximately two hours, and will keep me overnight for observation. In the morning they'll make sure the device is staying in place properly and the monitoring leads are all working. As long as there are no complications, I should be headed back home Wednesday afternoon or evening... though I hesitate to think things would go that smoothly -- reading back over the last few months, complications and the unexpected were par for the course.

I'll have a sling temporarily while the surgery area heals, and my range of motion for my left arm will be limited until they're satisfied the implant won't move or get dislodged. So for a while, both my left and right arms will be a bit unreliable -- so at least things are balancing out.

One of us will post another update after the surgery to let folks know how everything went.

Tuesday, August 23, 2011 3:23 PM, CDT
...and then the power went out

Likely due to some construction in the area for a new light-rail line, the University of Minnesota Medical Center lost power today while Jay was waiting for his procedure time.

Long story short, after enough delay his procedure was cancelled and rescheduled for tomorrow.

We'll try again.

> I did not think of it at the time, but thank God the power outage had not occurred the week of May 24th... I was hooked up to so many machines. In the time it would have taken for the backup generators to kick in – who knows what could have happened.

Wednesday, August 24, 2011 12:50 PM, CDT
The lights stayed on!

Procedure is done, Jay's getting settled in his room on 6C at U of M. He'll be here overnight and coming home tomorrow. Everything went fine, he's completely awake and everything.

Saturday, August 27, 2011 9:48 AM, CDT
In and Out and In and Out...

I'm home from surgery, recovering after the successful ICD implant surgery from earlier this week. My sense of time is all out of whack, as one of the prep medicines they administered before the procedure had amnesiac effects, as well as generally clouding up my mind.

I was a bit anxious going in, so they gave me something to calm my nerves, but it wasn't long after I arrived on Tuesday that I was wheeled into the operating room. Lots of people running around with tubes and wires and medical instruments. They had to completely shave my chest, and I still haven't been able to wash off the orange staining from all of the iodine used to sterilize the site.

What I wasn't expecting was to be awake during the procedure. For the implant part of the operation, there was a local anesthetic, and a small drape placed over my face so I couldn't see what they were doing -- since the surgery was taking place literally just inches from my line of sight. While I didn't experience any pain thanks to the meds, I could feel the pressure and some odd tugging and pulling over the course of the procedure. And of course, I could hear everything they were talking about.

Apparently, I wouldn't shut up during the procedure -- I kept asking questions about what was going on, how things were, if the leads were implanted yet, if the blood thinners were making it more difficult, if that feeling was them pulling the skin pocket open to insert the ICD, and if that funny smell was my own burning flesh while they cauterized the site (which it was -- the second time I've had to experience that. I smell a bit like baby back ribs).

After that, they told me they'd put me under completely so they could test the device. So off I drifted for some unknown amount of time.

I only vaguely remember coming to sometime later once everything was done and getting wheeled back to the recovery room with Trish and Mary (my mother-in-law). There was a bit of a snafu with my diet, where the kitchen had been told I was on a liquid-only diet, while the physician had cleared me for a normal diet. After a while that got all sorted out and I was able to finally eat -- fasting before the surgery and taking all my morning meds on an empty stomach had made me a bit cranky.

Then, most of the meds from the surgery started to wear off. It felt like I had been hit in the chest by a shotgun blast, or kicked right in the sternum by an angry mule. Wow.

It's the most intense pain I had experienced. It was also the first time I took a look down and saw the area they worked on. While it was covered with bandages, my left pectoral area was so swollen, it looked like they had grafted a Snickers bar under my skin.

The first couple medications they tried did nothing for the pain -- it kept racing up the pain rating chart as I started to clear more of those meds from my system. Finally, we settled on Percocet, which greatly helped reduce the pain, but has the unfortunate side effect of making me completely loopy and cotton-headed.

Strangely, despite the medications that made me all sleepy (Xanax, Ambien, and the Percocet) I slept horribly that night -- I kept startling myself awake like I had at the hospital after my heart attack, and was anxious about falling asleep and not waking back up. Every time the nurses came in to take my blood pressure, a blood sample, adjust an IV, or otherwise check on me, I was wide awake...

The next morning, though, several nurses stopped by to say hi. Apparently some of the staff who had cared for me during my rough first few weeks back in May & June heard I was having the ICD, so they wanted to come down and see how I was doing. That was a nice touch, and everyone seemed amazed and impressed at how well I had recovered, and mentioned that many of the staff were still asking after me and glad to hear I was doing well.

Thankfully, that weird paranoia/twitch-reflex I had been experiencing went away as the day went on, and I was discharged just after lunch. Driving home with Trish and getting to recover in my own bed instead of a hospital room sure made a big difference, and I slept pretty well, except for when the cat jumped up on my surgery site. Youchies.

Thursday and Friday both went by in mostly a blur. I've got some antibiotics and painkillers added to my regiment for the time being, but those should be wrapped up by next week. Until then, though, I'm pretty much in a deep mental fog after taking my meds, or in a brief moment of painful lucidity before the next batch is ready to go. Hopefully the swelling will go down by then, too, and the initial stiffness and soreness will fade.

With each passing day, I do feel a bit stronger, and more optimistic about my long term recovery. I'm looking forward to getting back into game design, spending more time with my kids, and being around more overall to take some of the pressure off Trish and hopefully start feeling like a normal family again. Well, except for the fact that I'm part cyborg... but other than that... :)

> And thus ends "The Event," as it would come to be known. I returned to the hospital and saw various doctors for several follow-up visits, more lab work, numerous tests, and even another trip to the ER on a false alarm. But this marks a certain turning point, when our outlook changed from trepidation and uncertainty to optimism and hope. So this was as good a place as any to move on. Both with the book and our lives.

7. On Nurses & Doctors

It's hard to overstate how important the medical staff – especially the nursing staff – is to your recovery after something so severe. Often, just by reading the message board on the wall and seeing which nurse was assigned to me, I would know if I was in for a good day or a bad day. They made that big a difference.

The good nurses tended to have the right blend of humor, compassion, optimism, and just a little bit of attitude – enough that you knew the nurse took his or her job seriously and that they weren't going to tolerate any shenanigans. This was fine by me. I didn't have enough energy for any hijinx, let alone shenanigans. The really special nurses were able to put me at ease, and feel empathy for what I was going through, without letting it interfere with their work.

And boy did they work hard. It amazes me how many people flitted into and out of my room on a given day. Early on during my stay at the Cardiac ICU, I had several IVs, a blood pressure cuff, oxygen saturation monitor, ECMO, PICC line, SWAN Ganz line, EKG leads, a heart monitor, and a catheter. Each of these needed tending throughout the day. Every day they also weighed me and took my temperature. I was on more than a dozen different medications. With everything they already had to do, I was reluctant to buzz the nurses for anything that wasn't pretty darn serious. I was already getting a lot of attention.

For as awesome as the really good nurses were, that's how bad the really bad ones were. To the point that some nurses actually scared me. They were rough, mechanical, distracted, or acted like I was taking time away from something more important. Not the sort of treatment you want from someone you depend on to keep you clean, fed, and properly medicated. Fortunately, there were only two or three nurses I had to ask the staff manager to never assign to me again – I simply could not mentally handle their demeanor or treatment. Seeing their name on the board for the day was soul crushing.

Credit Where Credit is Due
Rest assured, there were far, far more amazing, helpful, cheerful, inspiring, and downright nice people there to care for me along the way. I wish I could remember every single one of them. At the time, while I was constantly writing as part of physical therapy to recover use of my right hand, I kept making lists of the nurses – who I liked, who I disliked,

who would bring me ice chips, who had a good sense of humor. Things like that. I kept long lists of everyone and everything.

Somehow, in the crazy shuffle over the course of this medical odyssey, some of those lists were lost. I regret that I can't recall and properly thank all of them. But I do remember some of them. So here is a rambling list of some of the great people who helped care for me. I don't know if these people were at the University, Regency, Fairview, or elsewhere – I just know that they were amazing.

Just a few of the people who deserve recognition: Liza, Danielle, Tazim, Pat, Marilyn, Kim, Holly, Casey, Erika, Ashley, Cami, Kay, Genesis, Washington, Geneva, Lea, Eniola, Betty, Colleen, Hope, and Charla. I know there were more; I just can't remember them all.

Kim gets an extra nod because she was the one who told me I could refuse a medication or treatment if I felt uncomfortable with it, or even ask to not be seen by certain nurses. While I only did that once or twice regarding medication – until I was able to talk to a doctor or nurse who could tell me why a change had been made – it was empowering to be reminded that as the patient, I'm ultimately in control of my care. This also helped me make sure I had better care through my stay, since I was able to weed out some of the nurses whom I didn't like or have much respect for.

Geneva gets a round of applause for being the first person – after being at Regency for more than a week – to tell me that I could get juice, popsicles, or other small snacks in between meals or late at night when I couldn't sleep. And when she found out I liked those little Luigi's Italian Ices, she called down to the cafeteria and had them bring several up. She kept them in the nurses' freezer, labeled just for me. That was incredibly sweet and very much appreciated.

Kay gets a standing ovation for getting me scheduled for my first shower. She was the first one during my entire stay who asked me if I felt comfortable, clean, or needed anything hygiene-wise. It had been five or six weeks since I had showered… it was such a gross feeling to be stuck in bed like that. She also helped me with my first haircut, and got me an oral care kit – it had also been that long since I had brushed my teeth. Not something I'd recommend to anyone.

Lea gets special attention for really listening to me about my pain symptoms and not just waiting for someone else to do something. Some nurses would just give me a dose of this or that and move on, but when she read the previous night's notes that I was in pain, she followed up with me several times throughout the day, and got in touch with the resident on call and really got things moving to have a CT scan and find out that I had kidney stones.

Rock Star Treatment

There were a number of amazing doctors, as well. It's too bad I never even had the chance to meet the majority of them. Many of them were only involved during those early critical weeks when I was kept sedated. There were respiratory experts, the doctors who performed the surgery to attach and remove the ECMO, the cath lab specialists, and many more. There were so many specialists involved during those first few days. And I'll never know who some of them are. But thanks to all of you, whoever and wherever you are.

Among the unknown masses, though, there are two doctors who I do remember well. They had a significant impact on my recovery – and not just because my life was in their hands. The first doctor who stands out is Dr. Cindy Martin, the cardiologist who first saw me when I was transferred to the University of Minnesota and oversaw my initial critical care. Then later on, after I was transferred to Regency Hospital, another outstanding doctor, Dr. Hemant Trehan, oversaw my daily care.

Unfortunately, I don't remember much about Dr. Martin from that early timeframe – she was managing my care while I was mostly unconscious or under heavy sedation. And after reading over my medical records, I see that I kept her hands full. Later I was able to piece together that it was her face I saw and her voice I heard as I would drift in and out of consciousness – asking if I knew where I was, if I could nod or squeeze her hand, and so on.

Trish was eternally grateful for Dr. Martin, as well. Dr. Martin kept Trish updated in a candid, no-nonsense way that was still very humane and sensitive. Dr. Martin laid everything out plainly – the bad signs, the good signs, what they were hoping to see, and what potential challenges loomed on the horizon. It's exactly what Trish needed at the moment, in order to simply make sense of everything, as well as make some of the medical decisions necessary during the worst of it.

When I saw Dr. Martin a few weeks after discharge, I was absolutely thunderstruck. I only vaguely recalled seeing her while I was at the hospital. Seeing her face while fully lucid was like an end-of-movie montage that flashes through all the clues and hints that were sprinkled throughout the film – ending with the big "a ha!" moment when it all comes together; I think of it as the *Sixth Sense* moment. (So *that* was her..!) She's been my cardiologist ever since, and I completely trust her with my life – I already did once, and it turned out pretty good.

Then there was Dr. Trehan at Regency Hospital. Dr. Trehan was just… incredible. He was frank, candid, and one of the most genuine, sincere people I came across during my entire hospital stay. Or anywhere else for that matter. He was also very spiritual, and kept reminding me how miraculous my recovery has been, and how karma and life have a way of bringing us to where we need to be. But it was never preachy, and never seemed inappropriate.

In fact, he always seemed to know just what to say, and just how to say it. Dr. Trehan was the first doctor to go into the step-by-step details about what had actually happened to my body. He would pause occasionally to see if I wanted to know more or if I had any questions about what certain terms or conditions meant. I'm so glad he was the one to tell me. His demeanor and presence made it a lot easier to hear and start to process everything.

He also had an incredible outlook on life. Not only with a "pay it forward" sort of attitude, but a level of self-confidence that comes with accepting one's faults and failures. On one stressful morning, he told me how he had once been a very good medical student and a successful young doctor, but he was also an arrogant, self-absorbed asshole who was always miserable.

His candor surprised me, but it was refreshing to hear someone talk so openly. Somehow during that time, the tables had turned, and he gained a whole new perspective on life, coming to the realization that it just wasn't worth spending the energy to be anything but genuine, positive, and optimistic. It brightened my day and lifted my spirits whenever he stopped by. And it's been pretty good advice overall.

I'm confident that one reason my recovery went so well is I was in Dr. Martin's and Dr. Trehan's care. I can't imagine how things may have been different if another doctor had been assigned to my case at either the University or at Regency. Good thing I don't have to – I had the best care possible.

8. So What's With the Title?

It was a long time before I knew what had happened to me. What *really* happened. Early on, I hazily recall people mentioning a heart attack. And my mind had concocted dozens of other explanations, as well. I had drowned and washed up on shore somewhere. I had just passed out and hit my head. I had been the victim of kidney thieves. At first, I didn't know what to think. But Trish was there to walk me through it, slowly.

It's hard to relate to, and even harder to explain to someone who hasn't gone through something similar. It all seems surreal – like everything happened to someone else or some other family. Or you heard about it happening to a friend of a friend of a friend. So many events passed while I was unconscious, blissfully unaware of the trauma I was going through. Meanwhile, that was the same time Trish was in the most fear and under the most stress. She had to witness that trauma firsthand.

Trish had to see me in the room with all the machines and pumps and tubes. She heard the bad news and grim prognosis. She had to make critical decisions about my care. And she had to force herself to think about it over and over again so she could keep people informed on my progress. However, with each machine they removed or each test that I passed, she gained more hope, and the fear – while never truly going away – could start to slowly diminish.

Regaining consciousness was a real turning point for both of us. I think that is when our roles and perspectives on the entire event suddenly reversed. Once I was waking up with greater frequency and had longer periods of lucidity, my fears were just starting to develop. I was in a strange room with people I didn't recognize, tubes and wires hooked up to me, and with no idea what was going on. Initially, I couldn't remember anything. And I couldn't be sure if what I *thought* I had seen or heard earlier in the hospital was real or illusion.

As time passed and I gained more clarity, my fear kept increasing. The more I heard about what happened and started piecing together the puzzle of the event, the more I started to comprehend just how serious it was, and the more uneasy I became. It made me sick to my stomach. It freaked me out. Meanwhile, for Trish, the situation had been so dire while I was comatose that regaining consciousness was a blessing and a significant sign of improvement. While I sank deeper into fear, she was growing more and more hopeful the more lucid I became.

Trust Me, It's Coming

Ok, I admit, none of that tells you anything about the title. But it's context. I'm building the scene. Just wait for it.

Over time, as my senses and some memories started to piece themselves back together, Trish revealed more about the ordeal. She released details slowly to see how I would react. At first, most of the details were of the clinical variety – how did I get here? Was it really a heart attack? Did I really have complete renal failure? What do you mean I had pneumonia? I was under for how long? Where are we again?

Eventually, more substantial details emerged about the atmosphere, tension, and emotion surrounding the entire event. That's when I found out she was in the elevator on her way up to see me when she got the phone call on her cell that the doctors called a Code Blue. That's when I learned what Code Blue actually meant; I found out I had completely flat-lined, no pulse, nothing. I was being shocked back while Trish was coming out of the elevator.

Trish stayed out in the hallway, accompanied by the social worker and chaplain. There were more people buzzing about, but she can't remember who. She couldn't have been more than ten feet away from me. She said she could hear me roar in pain and anger each time the doctors had to shock me. And they had to shock me a lot.

After the team had stabilized me and prepped me for the cath lab, Trish was able to come to my bedside and see me briefly. There were a whole lot of people in the room, running back and forth. I was hooked up to countless machines, wires, and tubes. She recalls that I was still wearing my jean shorts. Amidst the chaos, before they wheeled me away, she kissed my hand. And then she leaned in and whispered, *"Come back to me."*

I heard her. Somehow, deep down inside, I know I heard her. I had to have heard… Otherwise, how else could I have mustered the strength to endure everything that happened next? I had a promise to keep. I had vowed I would be faithful and true, through the good and the bad, through sickness and health. And thankfully my body did everything possible so I could fulfill that promise.

Later, once we could look back on everything more casually – and even with a bit of humor – Trish admitted that in one odd moment, while it was all still so unreal, instead of saying *"Come back to me,"* she thought about saying, *"Come away from the light! Step away from the light!"* I'm glad she didn't. That would have made for a much more awkward book title.

9. Jedi Mind Tricks

The mind is a strange place. Especially when it has to keep itself entertained for a month or so. The more difficult part of recovery for me has not been physical rehabilitation, but mental certainty. I experienced memory loss. A loss of my sense of self. And my mind created literally dozens of delusions that it fed to me while I was comatose.

In a twisted way, a few of these delusions kinda' sorta' make sense, based on what was probably going on around me, or things my subconscious may have heard nurses and doctors talking about in my room. Others... well... Freud would have a field day. I have no idea where some of these originated.

It's hard to express how potent these delusions were at the time. No matter how bizarre, each and every one of these was *real*. Not a nightmare, not a daydream, but my actual life experience at that point in time. I lived each and every one of these from the fantastic to the frightening. Years of artificial life experiences and false memories were manufactured and packed into my head in the span of a few weeks. As the drugs started to wear off and my mind fought through the clouds for clarity, time and time again I would learn just how much of what I thought was true and real was nothing but a Jedi mind trick.

A few of these delusions are amusing, and several of them are even rather humorous. But there are also a lot of frightening, painful, and intense ones. This is a long chapter, and the amount of detail varies greatly from delusion to delusion, so I'm breaking the chapter up into smaller sections of somewhat related delusions – or at least delusions that came to mind or I think were in close proximity to each other. And these are just a small portion of the delusions bouncing around in my head. I made notes on many, many more than those included here.

A Day in the Life

These are some of the delusions that seemed to be based on something true – or at least plausible on some level. Or they were simply twists on what was really going on so the differences were subtle. These were the hardest delusions to be convinced were not true. I had a lot of trouble accepting that and letting go of some of them.

The Boil Order

I was always so terribly thirsty when brought out of sedation. When I asked for a glass of water, the nurses and doctors often had to tell me no. I didn't understand why I couldn't just have a glass of water. It was such a simple request. I didn't know what was going on around me. I would ask the nurses if the boil order was still in effect and if that was why they couldn't give me any water. They kept telling me they had no idea what I was talking about. That made me suspicious. How could they *not* know about the boil order?

Apparently, they didn't know about the pandemic, some virus that found its way into the water table. There were only a few water filtration facilities in the country equipped to properly treat the water to make it drinkable. Minneapolis happened to be one of them (for some reason, so were Denver and Austin). Until a permanent solution could be found, there was a permanent, government-instituted boil order active for any water consumed or used in a public or government facility.

I saw public service announcements playing on the television, reminding everyone "Boiled Water is Safe Water" or "Boil Before you Brush!" There were various news scrolls along the bottom of CNN giving updates on the progress of fitting more facilities with the proper filtration equipment. I swear the screen saver on the computer in my room and at the nurses' station flashed reminders about the boil order.

During this crisis, the USA was in close contact with China, working on some sort of trade agreement. Not only did China have the largest reserves of unspoiled water left on the planet, they were also manufacturing the machinery necessary to properly treat the water for its foreign partners. Each filtration facility in the world needed the parts China was manufacturing.

I remember seeing news footage of Barack Obama stepping off Air Force One after arriving in China. He walked down a long red carpet, where he was met by a short Chinese man with thick glasses and a wide smile. The man was accompanied by a geisha dancer. President Obama had just arrived for the big Worldwide Water Conference, where world leaders would decide how best to distribute water, manage the filtration equipment, and tackle the virus that started everything.

Infomercials

I watched a lot of television. Or maybe better put, a lot of television played in the background while I drifted in and out of consciousness. And for some reason, I love infomercials. I could totally have been an infomercial pitchman. There are some infomercials I've seen dozens of times and still never get tired of.

Well, there were two infomercials in particular that I recall watching – and I really did watch them. They were in constant rotation during my hospitalization. I was also completely convinced that I had already called in and ordered the products. And I was growing increasingly frustrated that they hadn't been delivered yet.

The first infomercial was for a health beverage called *Nopalea*. It's a red juice from the fruit of a rare South American cactus that has amazing healing powers. I watched so many testimonials about how it treated people's inflammation and joint pain, took away their back pain, the whole deal. It sounded like a miracle cure for all my aches and pains. But more importantly, it looked delicious. Ridiculously delicious. I had to watch as they poured the juice into a glass, in slow motion, while the 800 number flashed across the screen.

It was obviously the best tasting thing in the world – it had to be. It was so beautifully red. And everyone on the program loved it. It would make the perfect gift. My brother is health conscious and lives a holistic lifestyle. I was thoroughly and utterly convinced that I had ordered a "subscription" to *Nopalea* for him for his birthday. He would receive several 16 ounce bottles every month. I was looking forward to his next visit, because I hoped he would bring a bottle with him so I could try it.

The second infomercial had a lot less detail. It was for the new and improved Genie Bra. That infomercial seemed to be in a repeating loop it was on so often. What woman wouldn't want a more comfortable, better fitting bra? Why my wife would surely like one, maybe two, possibly three! I was convinced I had ordered several for her. The tan, black, and white, just to be sure. It's the best bra in the world, honey, you're going to love them. What do you mean they haven't arrived yet?

I can only imagine what the nurses thought, coming in to check on me as I sat there dazedly watching a commercial for bras.

Of course, it never occurred to me at the time, but a few weeks after overcoming these delusions, I realized I didn't have access to a phone or a credit card – how the hell had I ordered those things?

The Girl that Never Was

This one isn't fun. I can barely even type it. Just thinking about it hurts. My delusional family was very, very different from my real one. It tears me up inside to know that I didn't even recognize my own two sons the first time I saw them. I think I had to ask who they were.

The family I remembered was hopelessly grieving. In the delusion, Trish had several miscarriages, and we were starting to lose hope that we would ever have children. Then finally, almost nine months ago, she conceived, and made it all the way to the third trimester. Her due date had arrived, and then passed. We went to the hospital so they could induce labor.

We were overjoyed when we welcomed Rainbow Joy Little into the world. Our daughter was the most beautiful thing I had ever seen. Her little face would scrunch up into this tiny smile right before she was about to cry. Unfortunately, she was born with a severe birth defect – her lungs weren't fully developed or working properly. She was rushed off to the NICU and hooked up to machines to help her breathe.

We lost our daughter the next day, before either of us had even had the chance to hold her. Trish was distraught and inconsolable. I had lost my first and only child. And with it came the grim knowledge that this would change our marriage forever. It could very well drive us apart. We simply could not handle the grief.

Originally, one of the reasons I thought I was in the hospital was because I had fainted after watching Rainbow in the NICU and seeing her monitors flatline. Then being told by the doctors that she had died and they wish there was more they could have done.

It was the most horrible feeling I have ever experienced. And learning later that none of it was true was one of the most startling, frightening realizations I have ever had to deal with. I didn't just have a dream about having a daughter – *I had a daughter, and she died before my eyes*. She was taken away from me in the delusion, and then she was taken away from me again in reality. I lost her twice, and it hurt just as much the second time. I don't know how to explain it any other way.

There were a number of variations on this painful delusion. Sometimes there was a stained glass dragonfly window in our hospital room, and when the sun rose, it would cast a rainbow across the room, a constant reminder of what we lost. In some, Trish miscarried in the third trimester, but still had to give birth to our dead daughter. In all of them, our daughter died before either of us got to hold her. And in all of them, we had to bury a child.

I hated myself for so long. Why would my mind do something so cruel? This was incredibly hard to get over, because I simply could not be convinced it didn't happen. How could such a strong emotional bond be forged with a figment of imagination? Maybe Trish was lying about it. Maybe she thought she was protecting me by saying it never happened. Nearly a year after knowing she never existed, just thinking about my daughter makes me cry.

The Worst Care in the World

The worst person I was under the care of was a resident named Kim. He wasn't my doctor, but wasn't a nurse – somewhere in between. He worked the night shift, and monitored my care from dinner until the early nurse shift arrived.

Kim was Asian, and spoke with a slight stammering accent. He had gone to medical school in Japan (Osaka? Kyoto?) Amazingly, he went to the same school that my respiratory doctor had attended. Kim was awkward and uncomfortable to be around. He seemed far more comfortable with the lab work, machinery, and tests than with the actual patients.

Kim would come in and talk to me in the same dull drone, without ever making eye contact. Always making notes on his clipboard. He'd perform a series of perfunctory tests every hour – come in to check my pulse in my ankles, listen to my heart, check my O2 saturation, check all my IV rigs. Everything he did was very robotic and detached.

At midnight, Kim would unhook my IVs, take off my O2 sat monitor, and unplug me from the heart monitor. Then he would unlock the wheels of my hospital bed and wheel me into the corner of my room, to open up the space so he could set up his computer and have more room to work. Then before the first nursing shift began in the morning, he'd wheel me back to the middle of the room and re-attach everything before leaving.

Apparently I was still taking up too much space. The next night he opened up the sliding panel between my room and the next room. Then he pushed me as far into the corner of the adjoining room as he could. He pushed a number of various carts, wheelchairs, and gurneys around me, and stacked pillows up on them so he wouldn't have to hear or see me while he was working.

I remember being freezing cold, begging him for a blanket. Just one blanket. My teeth were chattering, my skin pimpled with gooseflesh. I lay there in the hospital, shaking and shivering uncontrollably. Kim would just turn his music up louder so he wouldn't have to hear me. I kept grabbing at the pillows around me, hoping I could pull some of them down onto me for warmth. Instead of falling toward me, the pillows fell away and onto the floor.

Enough pillows fell out of the way that I could look out into the hall and see the nursing station through the door. Standing in the doorway, silhouetted against the dim lights, was a petite female nurse with glasses and her hair pulled back into a tight ponytail. She made eye contact with me, and I mouthed "help me" over and over. She just shook her head, then pulled the drape across the doorway before walking away.

Back in reality, when a nurse would visit in the morning, I kept asking them to make sure Kim never saw me again, I wanted a different doctor, or nurse, or whatever he was. I never wanted him to come near me again. I'd try to tell them how he'd unhook me from everything and push me into the corner. They kept trying to reassure me that there wasn't anyone like that who worked there, but it took a while before I believed them.

Mott's Saline Juice

Like I mentioned earlier, I was always so thirsty when I'd come out of sedation. And while water sounded good, juice sounded even better. Especially since I was constantly being tormented by these small plastic 4 oz. tubes of Mott's Apple Juice. They were white on one side, with the bar code and nutritional facts, and had the green and yellow Mott's logo on the other side.

There were stacks of them over on the table. They were in the fridge. They were in the linen closet. The doctors had them in their pockets next to pens and stethoscopes. They were everywhere.

Whenever my saline drip would start to run low, a nurse would come in and grab one of those juice tubes. She'd draw the juice out with a syringe, then stick the syringe directly into my IV bag. I'd get just the faintest sweet, floral taste in my mouth whenever they did this. It made me so damn thirsty.

I couldn't understand why they wouldn't just give me the juice tubes. And why on earth make them 4 oz.? Who only drinks four fluid ounces at a time? I had visions of opening up dozens of them to fill up the pink plastic pitcher on my table.

I can't help but think the reason I kept having this delusion is that I could see them changing my catheter bag several times a day, and yep, that sure looked like apple juice. If apple juice was coming out, it must be going in, right?

Grape Slushies

Did I mention yet how thirsty I was? Like, all the time? Parched. And part of the frustration was that refreshment was just a few feet from me at all times, right there in the dispenser on the wall. In fact, why are there so many snack dispensers here in my room when you won't let me eat or drink anything!

I thought the latex glove boxes, hand sanitizer dispenser and other mundane objects in the room were snacks. I thought the hand sanitizer was dispensing whipped cream, not a foaming cleaning solution. The tissue box was filled with ice cream sandwiches. One of the drawers was full of popsicles.

But the most agonizing temptation was the big plastic purple container next to the sink. It looked like it had a flip top lid, and it was filled with grape slushy. That beautiful, purple grape slushy. I kept mumbling to Trish about how thirsty I was, and I was so mad at her for not giving me any of the grape slushy. It's… Right… There..! Just a sip… Just one spoonful!

Unfortunately, she had to break it to me gently that the big plastic purple container was not in fact a grape slushy dispenser but a container of antiseptic wipes. I think she just wanted to keep the grape slushy all for herself.

The iStore

Trish and I would sign up for any promotion, any sweepstakes, any chance to win a prize we could find. We always took the online surveys printed on your receipt for a chance to win some gift card. So it wasn't all that surprising that we won something. I was notified by email that we had won a gift card to the local iStore to buy some iKitsch. I wanted to surprise Trish so I didn't tell her.

Instead, I snuck out and drove to the iStore in secret. It was bright, gleaming white with bold colors accenting the store – kind of like the stark white and neon palette of the early iMacs. Everything was neat and orderly. Shelves were tidy and evenly spaced. Items were arranged equidistant from each other.

I remember that one wall had an enormous selection of colors to choose from. Similar to the paint sample color cards you can sort through, but I think these were plastic, showing the different colors and patterns the products were available in.

I was so excited. Trish was going to be totally surprised. I ordered a bunch of things for her. I got her a brand new 4G iPod. And then I got her an iCoffee maker, because I know how much she likes her morning coffee. And the iCrock pot would be really convenient to make healthy meals that could cook while we were at work. I'm pretty sure I got her something like an iCurling iron or iHair dryer, too…

Since there was such a huge variety of items and the different patterns and colors they could be customized with, you placed your order at a kiosk, similar to a wedding registry. I got a scanner gun and went through the store scanning the items I wanted, then loaded them into the kiosk. After completing my order, we could expect to receive our items in seven to ten business days.

But the gifts I purchased were never delivered. I kept calling to try to find out why. Finally, I drove down to the iStore to talk to the manager. Well, apparently I hadn't properly selected which patterns I wanted on the items when I scanned them. And by now, the gift certificate had expired, so without funds, they canceled my order.

Drowning, Strangulation and Suffocation

These were unfortunately three very common themes among many of my delusions and nightmares. I'm grateful that I don't remember the details from all of them – just small glimpses. But there are quite a few for which I still have an uncomfortably vivid recollection.

In a way, I can see why my mind would keep bringing up drowning, strangulation, and suffocation. My body was actually going through all those things. After all, I initially couldn't breathe on my own. I had things strapped to my face and laced around my neck, and I had developed pneumonia. It made a twisted sort of sense.

The Girl with the Dragon(fly) Tattoo... Maybe
One of the strangulation delusions didn't even start out that way. Rather, it started out as a tense murder mystery. It became the mystery of the tattooed woman.

Just days earlier, a woman had thrown herself off the tower of a local church. She had no identification on her, but she had a unique tattoo that covered part of her arm and back. I think it was either a hummingbird or dragonfly, but I can't remember which for sure.

Have no fear, Jay is on the case. I was going to solve the mystery behind her death and uncover her identity. In fact, I was forced to. The nursing staff refused to help me – no food, no changing the bed pan, no medication, nothing – until I started to solve the mystery.

For each clue I uncovered that they could verify, the nursing staff would give me one privilege or item back. The problem was, I couldn't leave the hospital and I couldn't use the phone. All I could do was watch an episode of CSI over and over that depicted the scene where the woman had fallen, and a shaky-cam home video of someone walking into the church and up the rickety stairs to the tower so I could see the path she took. I would fast forward, rewind, and pause, looking for clues.

At least the staff let me bring in witnesses. I was able to talk to the man who found her body and the policemen who first arrived at the scene. A young woman remembered seeing the victim arguing with someone at the back of the church earlier that day. It was a big, imposing figure, but was hidden by too much shadow and she couldn't remember much. Except he had a skull tattoo on the back of his hand.

I was able to use that information to find the man. He was a tattoo artist. When pressed, he admitted that he had an argument with the woman, but that they were good friends. In fact, he was the tattoo artist who had given her the hummingbird/dragonfly. For some reason, he wasn't the guy, but he did provide some other piece of information that kept the case moving forward. But now I have no idea what that information was.

The entire time I was investigating the mystery, I was trapped in my hospital bed. I had trouble moving my arms and legs, and the TV remote and nurses call button were always just barely out of reach.

I remember having the blankets and sheets wrapped around me so tightly, it felt like a boa constrictor looped around my body. The longer I took to provide new clues or evidence, the higher up my body the blankets twisted. Soon they worked their way up to my neck, and I could barely breathe let alone think or try to solve the mystery. And then I finally pieced the entire mystery together. I solved it. I had figured out both who the woman was and who murdered her by pushing her off the tower – just as everything faded to black and I lost consciousness, the blankets and sheets finally twisting around my head.

A Long Time Ago, in a Delusion Far, Far Away...
I was so excited that I was going to get a sneak peek at the upcoming Star Wars MMORPG computer game. In fact, I had been selected as one of the few beta testers. They were pulling out all the stops. I was flown out to LA for the beta test. They set me up in a swanky hotel. Then they told me to get ready for the beta test.

Ok, sure. Where's the computer? There's no computer. I was starting to get suspicious and wondered what was going on. Then gas billows into the room, and I black out. Next thing I know, I'm standing in a large, bland metal room which is completely submerged underwater. Some breathing apparatus is in my mouth, and I can see a tube running up out of the room, disappearing somewhere overhead.

Then Samuel L. Jackson's voice welcomes me to the Star Wars MMO beta test. They wanted to try something different, so they wouldn't be thought of as just another World of Warcraft clone. So they were going to replace all the characters from Star Wars with the X-Men.

I was in the loading screen, waiting to be assigned to a server. Then I was whooshed away to another room, which I remember being dark and dreary. It was muggy and the ground was sludge. I was on Dagobah. Apparently this was my starting zone. And of course, all of this is still underwater.

Samuel L. Jackson then asked if I wanted the tutorial, or if I wanted to skip it and dive right in. I took the tutorial, and was walked through Dagobah, entering several dark, slimy caves while learning how to open my inventory, or crawling through dank tunnels while Samuel L. Jackson told me how to equip items. Finally I broke through some underbrush and could see some light streaming in from above. Of course, all of this is still underwater.

It was getting harder and harder to breathe. I was trying to listen as Samuel L. Jackson was telling me how to open my character sheet or examine my skill tree. I kept skipping past these parts – anyone who has played an MMO knows all the basics. Then something he said about quests and experience points caught my attention.

Unlike traditional MMOs, where characters earn experience points for doing things like clicking on flowers or pretending to cut firewood, or fighting orcs, this new Star Wars / X-Men MMO was breaking the mold. With their revolutionary new technology, they had the resources to push the envelope.

There were no experience points. Instead, you earned air-time. That's why it was getting harder to breathe – while I was going through the tutorial, I wasn't out questing. Since I wasn't completing quests, I wasn't getting any more air!

But it doesn't stop there. This game is going to be revolutionary. Instead of having a boring level cap where the game essentially ends when you hit level 50 or something – and then all you have left to do is keep grinding away for the best gear and loot – they decided that the endgame content would be your release from the game and unlimited access to air. That would keep players motivated.

I don't really remember much else, but I think I turned around and saw Professor X, with a gold question mark hovering above his head. I rushed toward him to get my first quest and earn some air.

Violence, Revenge and Retribution

This represents another unwelcome trio of all-too-common themes. Many of my delusions were horrifically, graphically violent. These are the ones I will never be able to un-see. Whenever I start to drift off to sleep, these images are one of the things my mind stirs up to keep me awake. I won't go into much detail on these. I don't want to dredge all those images and memories back up any more than necessary. Plus, it doesn't make for a very happy read.

Even though I don't go into grisly detail, you may want to skip this part if you're squeamish. Here is an abbreviated list of a just a few of the graphic delusions.

Alligator Zombie Showdown

I was somehow involved in the making of a homemade zombie horror movie. For authenticity, we used real zombie alligators from the nearby toxic-waste-tainted marsh. The actors were never told about this. So when it came time to shoot the scary scenes when the alligators attacked, the actors were screaming in real terror and agony because they were literally being eaten alive by the zombie alligators.

Center Stage

There is this absolutely bizarre band from Finland called Lordi. They're sort of a cross between KISS, Alice Cooper, Rob Zombie, and Gwar. They're known for their over-the-top live shows, because the band plays in elaborate, detailed monster makeup and the performance uses outrageous special effects. I was a special guest on tour with the band. My job was simple. I was dissected live on stage during the intermission. I was gruesomely cut open with a collection of rusty tools, then taken back to the trailer and patched up just enough to keep me alive for the next stop on their world tour.

The Indian Massacre

My hospital wing was somehow on an Indian reservation. The tribe had been plotting a long, complex revenge plan against the hospital staff, who kept taking over more and more of their ancestral lands. The Indian tribal warriors set up an obstacle in the hallway. One of the nurses would have to move some sacred objects out of the way to get to her patients or something like that. Somehow that made all the following carnage legal, because the staff had defiled a sacred tradition written into the bylaws.

As soon as this breach of tradition occurred, we heard whooping and war cries as Indian braves burst from behind the curtains and spilled out of a Conestoga wagon that just happened to be there in the lobby for some reason. They were armed with spears, hatchets, and warclubs. Their faces were painted with white and red warpaint.

The Indian braves mercilessly butchered everyone they encountered. They were scalping their victims as they slowly, methodically moved through the building and down the hall. My room was at the end of the hall. I was in restraints, and had to lay there the whole time, unable to move. I had to witness the massacre, and watch as they slowly made their way toward me, knowing I would be next.

There were several variations on this delusion. Sometimes it took place during the Civil War and the butchering was done by either the Union or the Confederacy, depending on whose hospital we were in. There was also a version with vampires initiating the slaughter because the hospital had been built over a part of an ancient, cursed crypt.

The Zombie Apocalypse

The zombie apocalypse was upon us. At least, that's what we were being told. Some co-workers and I were trapped in the middle of a floor high up on some skyscraper. Being in one of the central rooms, we didn't have any windows to look out. But we did have a radio. Every station seemed to have the same emergency broadcast, warning of the zombies shambling through the streets, killing everyone they encountered.

But something didn't quite add up. The announcer wasn't a news anchor with a deep voice, but some slick, Johnny Fever style disc jockey. The DJ on the radio kept prompting us to stay put and wait for help. Don't go outside. Just stay right there.

When one of us would try to leave, the air would start getting sucked out of the room, or they would be jolted with electricity or suffer some other mishap. The DJ just kept telling us to remain in the room and everything would be fine. It would all be over soon.

He even called us by name and commented on what we were doing, chiding us for certain escape attempts. That's how we knew that somehow he was monitoring us from somewhere else inside the building.

I remember the room being decked out in a weird 50s retro diner car motif. The DJ kept announcing these fake events on the radio to keep us in place. A bridge had collapsed. There was a big car accident on the highway. Some big show had been canceled. So everyone should just stay put and wait for help to come.

Then there was a scraping sound from all the doors. A horde of zombies broke through the barriers and spilled into the room. And everyone was eaten alive in slow motion until everything faded to black.

Trial by Chocolate

Some delusions were just bizarre, non sequitur, and very confusing. I don't know how all the elements fit together, or how scenes transitioned from one to the next. And I don't know how I could have possibly thought something like this was real. But in this delusion, I was responsible for the death of a Polish child.

I was managing a large, industrial revolution style steampunk-candy workshop. There were huge furnaces, knobs, conveyer belts, cogs, gears and pistons everywhere. The place was like an anthill, there were so many workers bustling around in it. Amidst all the chaos, a young Polish boy had wandered into the factory and climbed up the rickety metal scaffolding to get a better view of the operations. He slipped and fell into a gigantic chocolate candy mold used for chocolate snowmen we made just for the holidays. When we heard a scream, we shut down the production line and rushed over to see what had happened. When we were finally able to pry the candy mold open with crowbars, we found a chocolate covered child's corpse inside. I just stood there, stupefied.

When the Polish authorities found out, they demanded revenge. It was my workshop, so it was my fault. I was told to get down on my knees and beg for my life while members of their police force would come up and put a Luger to my temple and kept playing Russian roulette or dry-firing their guns to intimidate me. Finally the Polish authorities convened and decided I would be tried using the same means by which the boy had been killed.

I was forced into an enormous man-sized mold for some sort of candy. Then the foreman pulled a handle, like throwing the switch for the electric chair, and scalding hot candy poured down into the mold from the hopper overhead. The mold was divided into three equal man-sized sections. The foreman had only released enough of the molten candy to fill two of the three sections.

If the section I was in was not filled in with candy, I would be pardoned and released. However, if my section filled with the molten candy, I would be found guilty as I was burned alive. I was found guilty. When they opened the huge mold, steam jetted out from the sides, and there I was, encased in hard candy, like Han Solo frozen in carbonite.

Ceiling Tiles

An inordinate number of my dreams and delusions prominently featured me looking up at square or rectangular ceiling tiles. This comes as no surprise, since for six to eight weeks I spent 90% of my life lying on my back, looking up at square or rectangular ceiling tiles. But ceiling tiles stood out even more in at least two of the delusions that I can recall.

The Kidney Thieves

When I first regained consciousness, I was dazed, confused and disoriented. I was constantly barraged with questions like "Do you know where you are?" or "Do you know how you got here?" Well, some of my initial memories for ending up in the hospital are very, very different from what actually happened.

I remember waking up in a dark room with flickering fluorescent lights and the criss-cross of ceiling tiles overhead. Everything looked grainy and was black and white like an old film. The very edges of my vision were dark and clouded. Every once in a while a face would appear from one of the dark edges, staring down at me. The ceiling tiles always stood out, white tiles in stark contrast to the black lines separating them.

Everyone's faces had white hospital masks over their mouths, but they all had large, dark eyes. They'd look back and forth among each other, hovering overhead, and were discussing something. I couldn't be sure what it was, they kept mumbling, and I strained to focus and listen.

Invariably, my vision would darken and I would drift off. Only to regain consciousness in much the same manner. Ceiling tiles. Fuzzy vision. Strange masked faces. Then one time I woke up and there was a terrible pain in my lower back, running along each side of my spine. The masked people were buzzing around the room hurriedly. I saw masked nurses carrying silver trays and scalpels, and some dark lumpy mass being taken away from me and placed in a cooler.

The pain had helped me fight through the fog, and I was able to finally make out what they were saying. The masked doctors were panicking. One was arguing they should continue to operate and take my liver and pancreas while they had me opened up. The other was arguing that they had run out of time, people would start to get suspicious. He pointed down at my hands and told the other, "He's got to be some sort of magician, or musician, or artist or something like that. Someone is going to know he's missing and start asking questions."

Apparently, the fact that my thumbnails were painted bright blue (which they were in real life) really had them worried. Surely only someone easily recognizable and quirky would paint their thumbnails like that. They were so worried about being caught that I was wheeled out of the room, and down a long corridor, watching rows and rows of ceiling tiles pass by.

After a series of metallic clicks, I felt myself being lifted up, and something loud slamming behind me. I was left in total darkness. I had just been hoisted up and thrown into the back of a van.

The next thing I remember was lying by the side of the road, the world back in Technicolor. I was in Arizona, lying on a patch of cracked, dry desert ground, looking up at a too-bright sun. Buzzards and vultures were circling overhead. I heard a screech, and an unfamiliar voice say, "Oh my god, honey – he's bleeding! We have to get him to a hospital."

I blacked out again, and the next time I regained consciousness, I was in the hospital, surrounded by strange people in masks, square ceiling tiles overhead. I initially panicked, but saw that everything was in color, and things weren't as blurry and grainy as before.

When I was asked if I knew what happened or how I got there, I thought I had been kidnapped by kidney thieves during a business trip of some sort, and dumped by the side of the road in Arizona. Good Samaritans had stopped by to help me and took me to the hospital. As strange as that was, I'm not sure it's any more unbelievable than what really happened to me.

But I still continue to paint my thumbnails. Just in case.

The Great Pizza Caper

This one is goofy, but I was taking it very seriously at the time. This delusion is less coherent than the others, and I only remember blips and short scenes or sequences, rather than the entire experience start to finish.

But telling you that it was essentially Scooby Doo and the gang working with me to solve a pizza caper in the hospital probably sets the right tone. You see, the hospital's annual pizza contest was coming up, but contestants were being scared off by some creature and there was growing concern the pizza contest would be canceled.

I got mixed up in the whole thing because I was one of the contestants. Each patient in the ICU got to enter a pizza into the contest. No, don't ask me how we were supposed to prepare or bake a pizza while in the ICU – apparently that wasn't important enough for my mind to worry about.

The winner of the pizza contest got some great prize. Part of me thinks it was release from the hospital and a clean bill of health. But part of me is equally convinced that we were all competing for an extra-large blue raspberry Slurpee. I've remembered it both ways.

I had the inside track and was the favorite to win this year's contest. I had actually been in the ICU so long that I had participated in the previous year's contest and knew what to expect. But with all the staff changes and cutbacks, nobody at the hospital was still around who had been there for last year's contest. I couldn't let the judges find that out or I'd be disqualified.

Then the memories and visions start to break up. I remember that somehow we made our pizza ahead of time, so we would have enough time to search the hospital for clues about who was trying to sabotage the pizza contest. My contest entry was this huge, rectangular pizza with lots of pepperoni. To catch the culprit, we devised one of those hair-brained Rube Goldberg-esque traps that appears at the end of each episode.

We ended up removing one of the ceiling tiles from my hospital room and replaced it with my pizza. Once in place, it was hard to tell the difference with just the crust showing. Now all we had to do was lure the suspect into my room, confront him, and then I'd hit my call button – which would drop the pizza from the ceiling onto the suspect.

Somehow, this would trap the suspect just in time to pull off his fake mask and reveal who the real culprit was. And he would have gotten away with it, too, if it weren't for that pesky pizza.

Business Savvy

Fortunately, not all the delusions were bad. The worst part of these delusions was how crushed I was they weren't true. I was so excited to get back to work. And thinking I was mega-rich wasn't bad, either.

Breyers vs. Blue Bunny
This was a long and complex delusion, because it was ongoing. It didn't span just one scene or one day – it was years and years of life as the founder of Breyer's ice cream. I remember how much pride I felt in owning Breyer's ice cream and building its success over the years – and felt our incredible desire to crush Blue Bunny and Edy's.

This entire delusion spanned the 1950s, World War II still a recent and painful memory. But the prospects of a space program loomed large in everyone's minds. We lived in Texas. I remember it was always so damn hot, and I wore a great big cowboy hat all the time.

Texas was one of the leading regions of the US where the government was looking to start the fledgling space program. The state lobbied long and hard, and I figured there was something I could do about it. I started to actively pursue the space program's first astronauts and technicians to endorse Breyer's ice cream and frozen novelties.

I even went so far as to create an entirely new frozen novelty department. They would be responsible for contacting the astronauts, developing exciting, flavorful new novelties, and coming up with a marketing campaign aimed at kids who were already daydreaming about one day walking among the stars.

We were able to secure a number of big name astronauts, but then we found out Blue Bunny was going after Buzz Aldrin. I was furious – had there been a leak within the company? We redoubled our efforts. I knew I could win Buzz Aldrin away from Blue Bunny if I could show him Breyer's was clearly the leader and had a solid market plan.

Our frozen novelty department developed dozens of new frozen treats with space themes. There were Lunar Punch Pops that were popsicles shaped like rocketships. We had ice cream bars enrobed in Crater Crunch – white chocolate with crispy rice cereal for a tasty, textural experience. We introduced a new line of ice creams and sherbets with names like Moon Grape, Saturn Swirl (which was kind of like Neapolitan, with narrower bands of different flavors packaged to look like Saturn's rings), Solar Sorbet, and Venus Vanilla – rich French vanilla again swirled with crispy rice cereal.

I remember travelling across Texas, and then across the entire country, in a huge freezer truck. I'd stop by diners and grocery stores to give out free samples and try to get them signed up to stock our new treats. If they did, they'd get a large cardboard point-of-purchase display of an astronaut. It even had the head cut out of the middle of his helmet, so a kid could walk up and stick his head through the hole and get a picture of himself as an astronaut.

It was slow at first, but really started to pick up momentum, and soon we were crushing the competition. I think Edy's ended up going out of business, and Blue Bunny decided to stick with just ice cream and non-dessert dairy products, giving up the frozen novelties part of their business altogether.

The Minnesota Twins

As awesome as it was to be the founder of Breyer's, it was even cooler being part owner of the Minnesota Twins baseball team. I absolutely love sports, especially baseball and football. We lived in St. Louis for more than a decade, and got to experience the St. Louis Cardinals firsthand, one of the most storied, popular franchises in league history.

When deciding to relocate for a new job, the available sports scene was actually important to me. The fact Minnesota had an NBA, MLB, NHL, and NFL team certainly worked in its favor. Unfortunately the Twins were in the American League instead of the National League. I guess you can't get everything you want. Well, not unless you're co-owner of the Twins. I lobbied for years with the league to allow the Twins to move to the National League Central.

Co-owning the Minnesota Twins with Trish was such a wonderful experience. I don't know how we had earned a comfortable enough living that we could afford the Twins, but that's irrelevant. Owning the Twins gave us the resources to really make a difference in the community. Trish was heavily involved in local charities, and getting the players and their families to volunteer and participate in community outreach programs. We started scholarship funds for students to come and attend school in the great state of Minnesota.

And we very nearly suffered the worst promotional debacle in league history. Free Bike Day. Attendance was flagging toward the end of the season, as the Twins' chances of reaching the post season started dwindling. Someone in marketing got the bright idea to have a Free Bike Day – every attendant 15 or younger would receive a brand new Schwinn bicycle as they exited the stadium.

Within 24 hours of announcing Free Bike Day, the game was sold out, standing room only, the most heavily attended and anticipated game of the year. But there was only one catch. The head of marketing had no plan to actually give away any bikes. He thought this was the best of both worlds – huge ticket sales and attendance, and no expenses to supply those silly bikes. The kids and fans would be upset for a while, but they'd forget all about it by the time next season rolled around.

I didn't find out about this until the seventh inning stretch of the game. Someone came up to our loge box and told me there were no bikes, and people were starting to panic. I was furious. I fired the entire marketing department on the spot, and had everyone attending the game with me follow me out of the stadium.

All of my guests and I took a dozen or so delivery trucks from the food services gates behind the stadium, and drove into town. Some of us went to Dicks' Sporting Goods, Sports Authority, Target, specialist bike shops – anywhere that might have bikes in stock. We bought everything we could, packed them into the trucks, and got back to the stadium in the middle of the ninth inning.

All of my bigwig guests, several of the management staff, and I stood at the different exits, supplied by one or two truckloads of brand new bikes. We made sure each and every attendee got a bike – not just the kids, but the adults, too. When the news media found out about the marketing fiasco, it became front page news. First, deriding the marketing department for their underhanded tactics. Then praising us for how the organization as a whole handled the situation. The following year, we had more fans buying season tickets than ever before.

And Much, Much More...

There are still dozens and dozens of other delusions and half-memories I have from my sedation and drug-addled time in the hospital. As you can already see, there's quite a lot of detail in these. And I know there's more detail there, but I can't put it into words. It's like looking at one of your children. You know every single detail of your child's face, but there's no way you could convey that information to someone else so they could visualize it perfectly in their mind.

So instead of going on for another twenty pages, I'm going to highlight some of the remaining delusions that stand out for some reason. Either its grip on my mind was pretty strong, or it had some shock value, or in retrospect it's just downright bonkers. This isn't even all of them, but it gives you a good idea how bizarre my life was at the time. Enjoy.

1. I was a soldier stationed in Osaka. We had to hold a Japanese bunker at the end of a long peninsula, protecting the shore from flying great white sharks. They were apparently great white supremacist sharks, as their flippers had fabric bands with swastikas on them.

2. I spent time as a park ranger. While making the rounds of nearby campsites, I came across a grisly scene. There were turned over coolers and a ransacked tent with a gore drenched sleeping bag. A feral bear had attacked the poor campers. Now I had to track down and shoot the bear. I tracked it back to its cave, following the trail of blood. Inside the cave, I found the bear already dead and partially eaten. A mad man was in the cave, covered in blood and filth, gnawing on a human arm. It wasn't the bear that had attacked the campers, it was this mad man. He tried attacking me, and I had no choice but to shoot him.

3. I was on trial for committing war crimes during World War II. I was facing a firing squad, when suddenly a single large red India rubber balls came bounding across the room. Somehow, that India rubber ball saved my life, but I have no idea how.

4. Another World War II dream. The Germans and Americans were on opposite sides of a gorge divided by steep slopes. We each had a number of big, dark blue metal luges/torpedoes. Each side would fill

their luges/torpedoes with as many soldiers as would fit, then launch them down the slopes toward the enemy's so they would crash. Apparently that was how this particular war was fought.

5. I attended an underwater extreme diving event. There were cages stationed around this underwater bar we were relaxing in. One cage had an octopus, another some sort of crazy looking deep sea fish. The event was tragically cancelled as one of the dancers wandered too closely to the cage with the electric jellyfish, and was subsequently wrapped up and stung-shocked to death. Yet another delusion where it was hard to breath and constantly felt like I was drowning.

6. From the privacy of my VIP lounge, which I apparently shared with half a dozen other silk-robe clad lotharios, we drank heavily and watched the Pay Per View event of the year. It was the Lumber Jack Monster Truck championship. Death defying stunts, like monster trucks jumping over giant buzzsaws, or navigating through an operational lumber mill. Our wine glasses were kept filled by scantily clad nymphs.

7. I watched the World Soccer Cup. It was the first time they were trying several new formats. First, there were three separate divisions. Next, each division would play its own unique style of soccer. One division had to play on ice with skates. One division had to play on a field littered with jungle gym style bars and obstacles. The third division was played on multi-level fields.
    The final game was a three-way match between the three division winners. It was played on a triangular pitch on a narrow peninsula. Waves would crash up from time to time, spilling over onto the field. Right before the half, a shark had surged out of the water along with one huge wave, and flopped around on the field until security could remove it.

8. I somehow found myself playing a boardgame – for my life! I had been tricked into playing an ancient Nordic boardgame made out of carved stones and bone by an old gypsy queen. If I won the game, I would get my soul back. If I lost, I was doomed to play the game forever and ever while she consumed my soul.
    We played for real stakes, as I had to actually fight trolls and harpies, navigate traps, and make my way to the end of the board to locate the hidden treasure chest. There was a variation on this where

it was a Civil War boardgame played in real life, where the pieces were Confederate and Union soldiers, who moved around the board and had to fight each other to the death to occupy a space.

9. I remember being invited to a swank underwater exotic fish game. Each space on the board was its own small coral reef. We moved around the pool-sized board, trying to collect the most exotic collection of fish. The player with the most exotic fish won. Spaces with hazards were truly hazardous, as a moray eel might bite you, or you'd be stung by a sea anemone.

10. There was a Halloween party at the local Sesame Street theme park. Everything was decorated in stark black and white. All the Muppets in the park were goth, wearing dark leather, had piercings, eye liner, the whole thing. Grover was white with dark black eyeliner and a small black tear tattooed below one eye. Big Bird carried a dead cat around with him. Elmo was now Emo, with long dark fur combed over his eyes so he didn't have to make eye contact with anyone.

    Music by the Cure was playing on the sound system in the park. I was in some security room somewhere on the facility, watching everything through the static and flickering lights of the security monitors. Kids were being snatched throughout the park, dragged off into the bushes by goth Muppets, never to be seen again.

11. To salvage their floundering attendance, the NHL and MLB decided to combine their leagues and try a joint season. Each franchise would have only one roster, and all the players on that roster would have to play both hockey and baseball. Half of each game was played on a baseball diamond, then they'd somehow switch it over to an ice rink to finish the game with the same players. Jim Thome ended up being a much better goalie than you might expect.

12. I was participating in the first annual Underwater Boardgaming Championship sponsored by BoardGameGeek.com's newest venture, WaterGameGeek.com. I was currently in the lead. Participants had a limited air supply, and could only earn more air by scoring victory points or winning games. As the competition heated up and more and more players were eliminated, one of the other players sent goons to harass me while playing – they would twist and bend my air tubes to cut off my flow of air, or try to stall me so my air would run out.

    The competition finally drew to a close with the two of us facing off in the finals, each gasping for air as we moved toward the end of the championship game. I don't remember who won. I had this delusion several times, and I think I played a different game during the final round in each of them.

13. My parents were so excited about the trip to Asia they had purchased for us as an anniversary gift. Unfortunately, the entire operation was a scam. After they checked us in, they showed Trish to her seat, and brought me to the back, where I was bound to a hospital bed and injected with a medicine that slowed down my metabolism and heart functions. Then they blackmailed my parents and Trish to revive me. "Because it just wouldn't be fair to all the other people waiting for help if we just resuscitated everyone."

    My parents paid the blackmail, and I was revived. We hurried to another airline to fly back to the states, and my wife was taken to her seat while I was kidnapped and medicated again, and the whole ordeal started over.

14. For some reason I was on a military submarine with a bunch of other civilians. We were on a top secret mission… somewhere. There we were, stuck underwater. Because of the mission we wouldn't be surfacing for months, possibly up to a year. Thankfully we had a supply of the ultimate cooking kit from some infomercial, so we were able to make k-rations in dozens of different ways, and it made "the best damn catfish you ever tasted."

15. I was hosting a lavish virtual reality cocktail party. There was a huge, luxurious outdoor pool with a waterfall and real sand beach. People were playing, sunning themselves and in general having a great time. Meanwhile, I was in a secret back room of the mansion where I was dealing with some pretty shady characters to program some VR files to produce the perfect wife. We figured it would make millions on the black market, as people could build whatever kind of mate they wanted – be it for companionship, love, sex, whatever.

    I wanted an assertive, aggressive woman. I ended up with a beautiful Asian woman with a dark secret. She was a sleeper assassin sent by a rival software company to see how much progress we had made with the VR programming. Eventually she was activated and told to take me out, and she chased me through this enormous forest for what seemed like days and days.

16. I had the privilege to attend a historic meeting. George Lucas and Gene Roddenberry were meeting to work on an accord and come up

with a mutually acceptable environment where everything from Star Wars *and* Star Trek could co-exist without any issues. Stephen Hawking was the arbitration official to make sure the talks remained on point and fair.

For some reason, I suddenly noticed this strange blue energy sheath connecting the participants. Toward the end of the session, we started wavering in and out of focus, images clouded with static. Apparently, the entire meeting was being projected from a satellite in deep space, and the four of us were having a holo-conference.

17. I had recently won Shoeless Joe Jackson's caramel machine from the 1900s during an auction. It was this enormous, blackened chunk of steel that looked like an oversized sewing machine. But it was essential to have finally gotten my hands on it. Because somehow, if I could prove its authenticity, it would allow me to finally solve the murder of this snobby white socialite.

    This was vitally important because the prime suspect in the case was my wife, a black maid whom no one would believe because of her race. I had to visit with her relatives, dig up old photos, search through old, tattered furniture, find some recipes we could make using the big cumbersome caramel machine, and for some reason knit these weird calico shawls. Yep, somehow doing all of that would prove her innocence beyond a reasonable doubt and we'd live happily ever after.

18. There was a huge Halloween candy collection for a charity party we were throwing. The hall was decorated with papier mache spiders, jack-o-lanterns, and bags of enormous candy... Not enormous bags of candy, but that each piece of candy was enormous. Like the size of a brick. We had to get everything sorted quickly, and get the candy stuffed inside the jack-o-lanterns before any of the guests arrived. And on closer inspection, the jack-o-lanterns were actually pumpkin shaped cakes.

19. I had found an old wooden music box in the dusty attic. I cleaned it off and opened it up, and found a tattered, yellowed sheet of music. I had found the original sheet music for some classical piece, and could use this to finally authenticate and prove who the composer was. This was more difficult than it seemed, as I had to navigate my way back through the attic.

    The only way to do that was to walk along this elaborate spider web of red silk ribbons. Each ribbon was woven through the other ribbons and eventually tied off to one of hundreds of different cracked, aged china dolls stacked along the shelves in the attic. If I wasn't careful, I'd step on a ribbon that wasn't supported well enough, and the attached china doll would fall from the shelf and crash, shattering on the floor. This would wake up the owner of the house. She would come up to the attic and yell at me, and I would have to start all over again.

20. There was a tall, attractive Asian nurse named Danielle who had smart black framed glasses and wore black scrubs. She was always encouraging me, and helped me regain the strength in my legs by bargaining with me. If I would get up and move to the chair for half an hour, I could have some ice chips. After her shift one day, I overheard someone talking about how Danielle had gone too far. The next day I found out she had been fired for "trying too hard to help me" by giving me ice chips when I wasn't supposed to get any.

    At least the entire delusion wasn't a lie. There really was a tall, attractive Asian nurse named Danielle. She even wore smart black framed glasses and had black scrubs. But apparently the rest of it is bollocks.

21. I was outside, enjoying a nature hike through a beautiful, lush forest. I came across a picturesque scene with a waterfall cascading down the mountainside into a boulder-dotted lake below. The water was a greenish blue and the edges of the boulders had a slight patina to them. It was a gorgeous scene, marred only by the dead animals littering the sandy beach, washed ashore by the waterfall's churning.

22. While I was in the ICU, a junkie was brought in. He was faking an overdose to get inside the hospital and try to steal some drugs. When a nurse found him sneaking around the medicine cabinets, she called security. The junkie forced everyone back with a dirty syringe. But finally he was disarmed, and taken away by security, riding out on the incredibly slow floor polishing machine that scrubbed the hallways in the middle of the night.

23. It was a dark and stormy night. The rain was beating mercilessly against the windows. I was growing increasingly frustrated with the nursing staff in the ICU. My recovery was not going as well as planned, and I was sick and tired of being told "no" all the time. No, I can't have ice chips. No, I can't have a glass of water. No, I can't watch ESPN (the only basic cable channel apparently unavailable in

the hospital). As far as I can tell, everything after the "dark and stormy night" part is actually true. What isn't true is that I ended up calling my dad and telling him to come pick me up at the hospital; I wanted to just go home.

I informed the nursing staff that my dad was coming to pick me up and I was checking myself out of the hospital. Since I was in charge of my own healthcare, they had to do what I said and help me get dressed and wheel me down to the lobby. But they just kept laughing at me and told me I wasn't going anywhere – which just made me madder and madder. Half an hour passed, then an hour… I kept calling my dad, imploring him to come to the hospital and pick me up, but he never did.

24. I was in a large, dark room with an enormous table in the middle. It was a war room. The table had a realistic map, complete with terrain features like real mountains and working rivers – essentially a miniaturized version of the region it depicted. There were several other men in the room with me. Most had scruffy beards, were missing teeth, and wore faded denim overalls and frayed flannel shirts. One was carrying a shotgun.

    Apparently we were in charge of the war being played out on the miniature map. There were hundreds of tiny people on the map, and we moved them around with long poles, positioning them for offensive maneuvers. Then we would just stand back and watch the miniature soldiers kill each other and laugh.

25. I was married to Felicia Day and we made out all the time. Okay, I just made that one up, but it would have been pretty awesome, delusion or not…

10. Facing Fear

As it turns out, dying was kind of traumatic. Go figure.

I still can't fully process what the doctors had to do to get my body functioning again. Reading my medical records was chilling, especially when I Googled some of the medical terms and equipment used. As I regained consciousness, and the magnitude of the event slowly started to dawn on me, the process of mental and emotional recovery came into play. Unfortunately, it couldn't start until I understood just enough to completely scare the [bleep] out of myself.

Fear was a constant companion in the hospital. I had frequent panic attacks, and would need to be medicated to help calm down. Everything was so strange, so unbelievable, so unrecognizable that it just fed my fears and anxieties. There were dozens of new people rotating in and out of my life. And so many cords and wires and things that beeped.

Made You Flinch

One of the most uncomfortable manifestations of my fear was my flinch reflex. It was especially strong those first few weeks after I woke up. Any time *anything* entered my peripheral vision I would flinch, cringing away from it. Everything startled me. Faces were still mostly blurry, or looked like the faces that haunted me in my delusions. My mind could not process the images and information coming in fast enough to let me know not to worry.

Each time a nurse or doctor came in to check on me, or a monitor would start beeping, or my blood pressure cuff would fire, or a high contrast image flitted across the television, or someone would walk past my room in the hallway… It would all cause me to flinch. All of them kept happening dozens of times each day, as part of the staff's normal routine. This took a long time to push past.

To Sleep, Perchance to Scream

During most of recovery, even after I was transferred out of the ICU, I was terrified of falling asleep. Even closing my eyes for too long scared me. Every time I closed my eyes, terrible, violent images from my delusions appeared. I couldn't even shut my eyes for a few moments to rest. I feared that if I closed my eyes and fell asleep, I'd slip back into a coma, or never wake up again. Or worse, slip into one of my frightening delusions where I was being suffocated, killed or mutilated, and be stuck there forever.

My body and mind fought against sleep as hard as they fought to survive those first few weeks. Whenever I would start to drift off to sleep, or my eyes would flutter shut, I would startle myself awake with a quick jerk. I was instantly alert and paranoid as a flash of fear-inspired adrenaline rushed through me.

It wasn't uncommon for nurses to come in at midnight or three in the morning for blood work or a test, and I would still be up. Initially, it felt like I never slept, I was always fatigued, always exhausted. I had to be given sedatives to force me to sleep so my body had time to heal. It is not pleasant to be forced to drift off when your mind is screaming at you to wake up, terrified that you're being forced back into a coma.

Sleep continues to be a problem, even now. I had insomnia before any of these events happened. That worsened tenfold. I have sadly reached the point where I cannot fall asleep without medication. I've tried, but I just can't seem to shut my mind off.

There have been plenty of nights where I'm still up at midnight, one, two, or later – even when trying to go to bed early. Or, there have been evenings where I would finally fall asleep, succumbing to the fatigue, but then something would startle me awake at two or three in the morning, and my mind wouldn't let me go back to sleep.

My doctor had to keep increasing the dosages of my sleep medications, because I had built up a tolerance to the previous dosage. There was a point when even after taking melatonin, Temazepam, and several Xanax, I simply stopped feeling tired. Now, I routinely take four to six pills at night, just to be able to fall asleep. That's in addition to my normal medications for the heart condition.

Dying a Little Each Day

The last lingering fear is one I continue to face, and wonder how I will ever overcome… or if I even can. It's the nagging fear that I'm going to die soon. It seems like I feel that way every day. The fear of imminent death doesn't hang overhead every second, but it is there, lurking. It is hard to remember if I have had a single day since my release that I have not felt at least one symptom associated with a heart attack, or felt the anxiety of imminent death. And everything around me just stops.

The paranoia and uncertainty return. Many of the symptoms of a heart attack are fairly common on their own, or symptoms associated with a number of less severe conditions. But with my history, I always fear the worst. I have frequent pain in my left arm and arm pit. I have hot and cold flashes. I might have a sudden wave of nausea or sweating roll over me. I get light-headed and dizzy very easily – almost every time I change elevation, such as getting up from bed or getting out of a car.

But there are mundane causes for many of these symptoms. For example, I still haven't fully recovered from the nerve damage I suffered, and my doctors seem convinced my left arm pain is not cardiac-related but neurologically-related. But it's still a left arm pain that feels similar to the pain I experienced before both heart attacks.

And the light headedness makes perfect sense, considering how much lower my blood pressure is now than it was before, how much weight I've lost, and the mix of medications I take every day. More than half of my prescription bottles warn of dizziness as a possible side effect.

I know there are a number of perfectly reasonable explanations. But I keep waiting for another symptom to manifest. I keep wondering to myself, *"What has to happen for this to escalate and I'm going to go to the ER or call 911?"*

Even though the rational part of my mind tries to tell me not to panic, it can't convince the rest of my mind. If I have two or more symptoms within half an hour, I usually start to freak out and might have a panic attack. There's always that one tiny bit of uncertainty. That one sliver of conviction that says, *"This is it, this is the one."*

It's physically and emotionally draining to live under a cloud of fear like that. I've sought counseling, seen psychologists and psychiatrists, been reassured by my cardiologist and others. And I've survived for months and months feeling these symptoms and having it turn out just fine. I've worked out, lost weight, and changed my diet.

None of that matters, though. Those professionals aren't feeling the symptoms. They aren't immediately reminded of the worst case scenarios that followed. Their minds aren't flooded with the images of violence, mutilation and horror that I experienced while I was unconscious. They don't suddenly feel like they're being suffocated.

But I do. All of that rushes through me each time I feel a symptom. Sometimes the symptoms and resulting fear are more severe than others. That's no comfort, though. It's incredibly hard to explain it, and I don't know if I've been able to do that. But nearly every day, there is a point at which I think I am about to die.

And that makes it very hard to live.

11. Taken for Granted

The old saying is true. You don't know what you've got until it's gone. Only by going without some of the most mundane things, did I really start to appreciate how lucky we are to live where we do. It is incredibly easy to take things for granted – until they're not there anymore. Even the simplest pleasures in life suddenly aren't so simple.

Here are some of the things I realized I had taken for granted, and came to better appreciate over the course of my hospitalization and subsequent release. Hopefully I can maintain this perspective as time passes, and appreciate just how fortunate we really are.

Sunlight

It is so easy to forget how direct sunlight feels so much *fresher* than artificial light. I was under fluorescent lights for so long, the first few times I had the opportunity to go outside and feel actual, honest-to-God sunshine on my face was rather startling. I forgot how warm and tingly it made your cheeks, or how after you closed your eyes, there would still be little dots floating around your vision.

Feeding Yourself

Being on a feeding tube for so long, I wasn't sure what to expect the first time a tray of food was brought to me. I needed to use specially weighted, oversized utensils because I was having so much difficulty feeling and moving my hands.

That first meal took forever. I couldn't cut anything. My hands just fumbled with the utensils and I kept dropping them. Even when I could get food onto a fork or spoon, my hands shook so much that I'd constantly spill food all over myself.

I hadn't used my jaw in so long it actually hurt to chew. I was growing extremely frustrated. But then I dug in my heels. *This is my dinner, dammit. And I don't care if I only get two or three bites of food, I'm going to do this.* I wasn't going to let myself feel like a toddler and have someone else spoon feed me.

Water

Oh my God water tastes so good. Cold, crisp, clear, clean water. We're spoiled. As Americans, we don't have to worry about this essential staple of life the way so many people around the rest of the world do. Turn on any tap or faucet, and you have more water than you could possibly ever drink. When I was first given water, it was almost overwhelming. It tasted so good. It was so refreshing, I had to sip it slowly, allowing myself to savor every drop.

Personal Hygiene

Being completely at someone else's mercy for your own cleanliness and hygiene can be tough. Plus, it's hard not to nitpick how they're doing it wrong, even though you can't do it any better yourself at the time.

It was a great feeling of accomplishment to be able to brush my own teeth, shave, use a Q-tip, and other simple staples of personal hygiene. The bigger steps were being able to go to the restroom on my own, and being able to shower by myself. It's hard to express what a feeling of accomplishment those two milestones represented. Huge.

Hugs

Imagine lying on your back, only slightly elevated, hooked up to countless monitors, wires and machines. It's hard to feel human when your physical body cannot support itself. In that prone position, the lack of human touch creates some incredible loneliness. Even just a hand on my arm when taking my blood pressure, or a pat on the shoulder when going over a test would help. Some level of human, skin-to-skin contact.

But what I really wanted more than anything was a hug. Partly for the emotional comfort it would provide, but just as much for the tactile sensation of a warm embrace and contact with another human. But hugs are impossible when you can't move your arms, rotate, and are stuck in a completely prone position.

When I was able to finally sit up and had enough wires and tubes removed that I had more mobility, Trish and I shared a good, long hug. I probably was squeezing as hard as I could, which wasn't much. It was one of the best hugs in the long and storied history of hugging.

Your Livelihood

Whether we'd like to admit it or not, what we do for a living says a lot about us. It defines part of who and what we are. For me, being a game designer isn't just a job or a career choice. It's a livelihood. It invigorates my life, provides a purpose, generates a legacy, and most importantly, it is a positive outlet for my outbursts of creative energy.

There was a moment in time when I was terrified by the prospect of never being able to return to work – not in the way I knew I had once been capable of. I was still foggy, I couldn't concentrate, and my hands were all but useless. There were nights when I'd lie awake wondering, what if I've lost that spark? What if the ideas don't come back? What if I can't design games anymore? Who am I without that?

Scheduling

Part of being at the mercy of others is giving up your control over your schedule and regular routines. It doesn't matter if you usually don't eat breakfast before 9AM. You better order it between 5:30 and 8:00 or you're waiting until lunch.

Different medical procedures and tests were scheduled throughout the day, seemingly at random. I'd be woken up in the middle of the night for blood draws or to administer a shot. I'd have to go to the bathroom really, really bad and need to wait 10-15 minutes for a nurse to come after hitting the call button.

Sometimes schedules conflicted – like being scheduled for a physical therapy appointment at the same time the x-ray technician shows up to take some x-rays. Or having the doctor making rounds and interrupting your current rehab session, but taking so much time that the therapist can't finish your session and has to move on.

Regaining control over my schedule was a big step toward re-asserting my independence. Eating when I want to. Trying to fall asleep when I want to. Using the bathroom when I want to. Rolling over in bed when I want to. Even writing a book when I want to.

Sleep

I'd like to say I've taken sleep for granted, but since I sleep so poorly and irregularly even now, it's more honest to say I wish I knew what it was like to take sleep for granted. That would at least assume I've slept well at some point.

Shifting Perspective

Initially I could not do anything on my own. I couldn't pull my sheets up if they got tangled. I couldn't adjust my own pillow. I couldn't roll over on my own. And I couldn't just swing my legs over the side and sit up. Being able to change my position was significant – not only to feel the physical progress of recovering my strength, but to actually shift my eye level and perspective.

I had been stuck in that one position, at that one angle, staring off into space in that one direction for so damn long. I wanted to be able to sit up and look around. Look out the window. Look out into the hall. Even look at the floor. Being able to sit up marked a major milestone in recovering control over my environment and had a much greater impact than I expected.

Swallowing Pills

With all the medication I'm on, how could swallowing my pills possibly be something I took for granted or missed? Well, let me tell you what happens when you can't swallow your pills. They find a way to force them in, through one end or the other. One of my least pleasant memories was morning and evening medication time early on while I was still in the ICU. I was on the feeding tube and couldn't swallow anything yet, so the nurses had to administer my medication via syringe.

This wasn't one of those tiny syringes with a little needle. It was more like a turkey baster. It was an industrial sized, pump action shotgun of medical injection. But they can't just fire off solid pills into your stomach. I don't know if it was because my stomach couldn't digest them properly, or it would do too much damage to the lining without some food to soften the blow. But for whatever reason, they had to be injected with water using that huge syringe.

The nurses would take the dozen or so pills I was scheduled for, and grind them up using this big, clunky machine that looked like a cross between a paper cutter and a stapler – and it had a really ominous name, like the *Knight Dream Crusher* or something. I could hear the *crunch crunch crunch* as the pills were pulverized into powder. Then the nurse would pour the powder into the syringe and top it off with water before putting the stopper in. A few shakes to try and mix the powder up into the water, then they would plug the syringe into my feeding tube and push. It would come out directly into my stomach in violent spurts.

I hated that feeling as the medicine whooshed into my stomach. And it often took three or four injections to finally get all my pills in. Not only did it taste awful to have those crushed, uncoated pills swishing around, it hurt like crazy. My stomach felt more distended and cramped with each injection. Once the cramps subsided, then I got to experience the nauseated feeling of having all those medications hitting an essentially empty stomach at the same time.

So that, dear friends, is why I missed being able to swallow pills.

12. Dignity is a Lie

It's hard to retain any sense of dignity when you know how independent you used to be, but now you can't even wipe your own ass, and you have to pee in a bag. And after a while it's hard to feel shy or embarrassed when you're wearing nothing but a hospital gown for nearly two months, and you've got people constantly buzzing in and out of your room. Nobody has second thoughts about lifting up your gown to check a wound site, inspect a suture, check for a rash, whatever. Even when the door is open.

Enough said. Moving on now.

13. If I Hear "You're Too Young For This" One More Time

You hear a lot of things while you're in the hospital. Medical jargon and acronyms as doctors discuss things with each other. Casual conversations between nurses as shifts change. The background buzz of dozens of electronic devices. Beeping, whooshing, and hissing from all the machines and monitors.

I had my share of interesting earfuls during my hospitalization. Some of my nurses were... how should I say... very colorful. Some told bawdy jokes. Some shared things overly personal. Some would talk to you like you couldn't really hear them or you were a child.

But there were a few things I heard that left an impression on me. These things tended to stick with me either because of their magnitude or their frequency. Here are just some of the more interesting things I remember hearing.

Can you hear me?
This is the first actual, verified memory I have since crashing in the ER in May. I was drifting through my delusions until suddenly this woman's face with dark curly hair appeared, literally parting the clouds overhead in my mind. She was staring at me, and her gaze was bright and intense like a spotlight. She kept saying, *"Can you hear me?"* or *"Can you squeeze my hand? Come on Jay, I know you can do it!"*

That was the first time I met Dr. Martin, during one of the many times they tried to pull me out of sedation to see how well I could tolerate breathing on my own. I have no idea when during that time period this happened, but as far as I know, these are the first words from reality that broke through the fog.

Do you know where you are?
This question was asked a lot. Each time they pulled me out of sedation, and quite a few times even after that. It was just one of a series of questions a nurse had to ask at the beginning of each shift. I heard those questions so many times that eventually, when a nurse would enter, I'd just put up my hand and say, "Minnesota, January 2nd 1973, Barack Obama, July, Yes, No, No, Yes."

Can you hit a nail with a hammer?
This was one of the series of repetitive questions I was referring to. They didn't think I fared well the first time I was asked, but fortunately Trish was there, and was able to smile and say, "He's just being a smartass."

Apparently, I wasn't able to talk yet and still intubated, so when they asked if I could hit a nail with a hammer, I looked dazedly around for a bit. I pointed to myself and shook my head no, then pointed to Trish and shrugged my shoulders as if to say, "Maybe she can."

Your recovery has been miraculous
I heard this proclamation, or variations on the theme, dozens if not hundreds of times over the course of my stay. This was probably the second or third most common thing said around me. It's one thing to hear this from an aide or a family member or a friend… but entirely different when you hear it coming from all the different doctors and nurses involved in my recovery.

I didn't know just how true this was, or why so many people kept saying it, until I read my medical records. It really was a miracle. Or more like a series of back-to-back-to-back miracles.

You were pretty much dead when we got you
That was the first thing said that left me physically dazed. Jaw dropped, eyes wide, dumbfounded, the whole deal. It confirmed how serious Trish had been trying to tell me things were, but as subtle as a jackhammer. No beating around the bush. No punches pulled. The "d" word. Let's just get that out in the open and move forward from there.

I don't know exactly when this was said, but I'm pretty sure I didn't have all the sedatives completely purged from my system yet. And it wasn't some EMT or resident who said it. It was my cardiologist from the U of M. I was shocked to see that Trish *wasn't* shocked by this. She had known all along. I can't imagine what that must have felt like.

No, you can't have that
Until my breathing tubes and trach came out, and I was able to pass my second swallow test, this was probably the most common thing nurses said to me. I was so thirsty. Always thirsty. I'd ask for juice or water, just a sip. How about some ice chips. Surely I can have a few ice chips. But they had to keep telling me no. Very demoralizing.

We never had a daughter
This was a real shocker. I outlined the full details in Chapter 9 on delusions, but being told that Rainbow Joy Little did not exist – and

never did —was a punch to the gut. Of all the delusions, this was the most real, because the grief of losing her was the most real. I just couldn't accept it. I think a part of me still hasn't.

All things considered

This became the standard caveat during virtually every exchange with nurses and doctors. Everyone appended it to anything said regarding any of my symptoms, changes, progress, or recovery. When you think about it, the "all things considered" caveat basically means "compared to being dead, yes, this is much better." Here are some examples.

> **Nurse:** How are you feeling today, Jay?
> **Jay:** I feel pretty good, all things considered.
> **Actual Sentiment:** I feel like hell, haven't slept a wink, but at least I am not completely freaking out at this very moment. Oh, and everything hurts, but at least it hurts *slightly* less than last night. And may I please have an extra blanket?

> **Jay:** So how did my EKG come out?
> **Doctor:** Not bad, all things considered.
> **Actual Sentiment:** Considering your baseline is way off from a normal person's baseline, and accounting for how completely fried your entire physiology is right now, I have no bloody clue. But at least I know the machine's paper roll was properly loaded.

> **Nurse:** How was dinner?
> **Jay:** Pretty good, all things considered.
> **Actual Sentiment:** It was dreadful. The food was bland, tough, and smelled funny. But since this is all I've had to eat in six weeks and I can kill the taste if I sprinkle enough Mrs. Dash over it, I ate like a king.

Just try to relax

Why thank you nurse. That is just what I needed to hear when I am totally freaking out right now because more waves of hot and cold flashes are washing over me. Or my right arm is throbbing and feels like nails are being driven into my flesh. Or I've startled myself awake and I have no friggin' clue where I am or what has happened. Or I think you're trying to steal my kidneys. Just try to relax. It's so simple! Why didn't I #@$!ng think of that?

Your wife is amazing
I kept hearing this from the staff over and over. And they're absolutely right. Yes, Trish was there by my side all the time, and she walked me through the events as I recovered. But I never really pieced together just how much she endured, and how strong she had to be, until I was able to read the CaringBridge account for myself.

It was moving to read about how she was able to manage the finances, career, home, work, kids, and still be there for me every day. It was amazing to keep a calm, controlled exterior while emotional turmoil swirled inside. For as often as it was said, it was still never said enough.

I have never seen anything like this in my life
It wasn't uncommon to hear remarks about how unusual or novel my medical experience was. But hearing that from a doctor is something else. Some said they'd never encountered a case like this in their thirty-odd years of practice. Or that they'd never even *heard* of such a case from a colleague. Maybe in someone with more risk factors, or in their seventies or eighties, but you?

Hearing this triggered a strange reaction from me – part pride for surviving it and being so unique, part creepy for being that unique, and maybe a little bit nervous wondering just how well they can treat me if my situation is that much unlike anything else they've encountered.

Your case is the holy grail
I heard this several months after my release. I was having some persistent pain in my arm, so I was sent to a neurologist for an EMG test. The doctor who administered the EMG had been running late after tending to an emergency, and hadn't had time to read over my entire medical history beforehand.

He ended up reading it while administering the tests, and was just stunned. That's a reaction I've seen a lot. But then after reading through it and asking a few questions, he said something that I found really amusing – that my medical case was like "the holy grail of case studies" for someone's research project or doctoral thesis. I smiled and told him I hope they get an A.

You're too young for this
Hands down, I heard this more than anything else. Ever. I wouldn't be surprised to learn that administration had sent out a memo making it mandatory to say this the first time someone came to visit me, because that's what it felt like. Doctor, nurse, attendant, X-ray technician, EMT, ambulance driver, respiratory therapist, rehab instructor, psychologist,

psychiatrist, social services liaison, pastor, cleaning woman, food services guy. It didn't matter. When someone entered my room or met me for the first time, it felt like this was the first thing out of their mouth.

It was kind of cute for a while, especially when one of the young pretty nurses would say it while patting my hand comfortingly. But it definitely started to grate on my nerves after the first few hundred times I heard it. I began to feel really snarky in my mind, and kept coming up with responses I am so glad I didn't end up using... It's never a good idea to make enemies with the people who handle your food, your medication, or your personal hygiene.

Here are just a few of my unused retorts for "You're too young for this."

- Oh really? I hadn't noticed.
- Thanks! That really cleared my arteries right up!
- Why didn't you tell me that *before* the heart attack?
- I'll try to do better next time.
- But I was too old for rickets.
- I didn't realize we were on a schedule.
- Sorry for the inconvenience.
- Well you're too old for whatever the hell it is you're doing.
- I'm glad you mentioned that, because nobody else has brought that up before. Ever. And by "ever" I mean *all the freaking time*.
- I'm not too young to kick you if you were any closer. And if I could actually use my legs right now. Because if I could, you'd totally have to watch out for that.

Thankfully Trish gave me the tools to handle it, courtesy of the awesome cartoon *Phineas and Ferb* on Disney Channel. These two brothers are always creating crazy inventions and cooking up impossible schemes to help pass time during summer vacation.

Invariably, they'd run into someone suspicious of their shenanigans. Like trying to buy enough steel to build a roller coaster, or buying a dozen steer for a rodeo, or driving a forklift. Whenever the situation came up, one of the adults would look at Phineas and say, "Aren't you a little too young for [whatever scheme you're planning]?" And Phineas would just look at them, smile and say, "Yes. Yes I am." That became my mantra.

14. Tools of Recovery

Recovery is a long and arduous process. It can be frustrating when you have to take baby steps when you know you used to be able to run. And recovery comes in a wide variety of forms. Physical, mental, and emotional recovery are all equally important after a traumatic experience, but tend to progress at very different rates. I was fortunate enough to have some of the right people and tools available to help me with all the types of recovery I needed to truly get better.

Planning Ahead

A big part of the entire recovery process was the preparation itself. Making a plan. Getting into the right mindset. Taking the time to stop and think through what you were trying to accomplish, and being able to set and evaluate realistic goals. Sometimes this was the toughest step, but I'd find that once I had the right plan in place, following through with it came naturally, or was much easier than I had expected.

One example of planning ahead was when I learned about the upcoming swallow test after my trach tube would be removed. I knew I wanted to pass and start eating real food again. But not just any food. I knew there were some foods I really wanted to taste again, but this also got me thinking longer-term about changes to my diet and portion control after my discharge.

I started making lists of the types of food I wanted to eat. The ones I was really craving. Then crossing off the items on the list I shouldn't eat any more. Early on, the list was pretty simple. Rainbow sherbet, strawberries, Jell-O, orange juice, and yogurt were at the top of the list. And ice chips.

It's ridiculous how much I craved ice chips. Ice chips were always so tantalizing. No real taste, but so cold, such a different texture, and slowly melting to help slake my thirst. I would make notes about which nurses would give me ice chips and which ones wouldn't. And I wanted some Gatorade to pour over a cup of ice chips so I could make my own homemade Slurpee. Mmmm.

But ice chips served an important purpose. I had to be patient, take my time, and methodically force myself to practice slurping and swallowing. As silly as that may sound, my mouth, tongue, jaw and throat had been out of practice for over a month, and I'm glad I had thought ahead to practice on the easy stuff so the normal food was easier to adjust to.

As I mentioned, one part of planning ahead was looking at our diet, food selections, and meal planning. This is one thing I was able to get the kids involved with. I had them help me brainstorm the three most important things about the types of food I would want to eat going forward. After some discussion, we came up with Nutritious, Delicious, and Convenient. If a food wasn't all three of these things, I should re-evaluate how much I wanted it.

Ben got really excited and told me about these "bubble charts" they had been working on at school. He drew one out on some paper for me, and I realized he was making a Venn Diagram of those three characteristics. We pretended we were at the grocery store and thought of a whole bunch of different foods and meals we could make, and figured out where to plot them on the diagram.

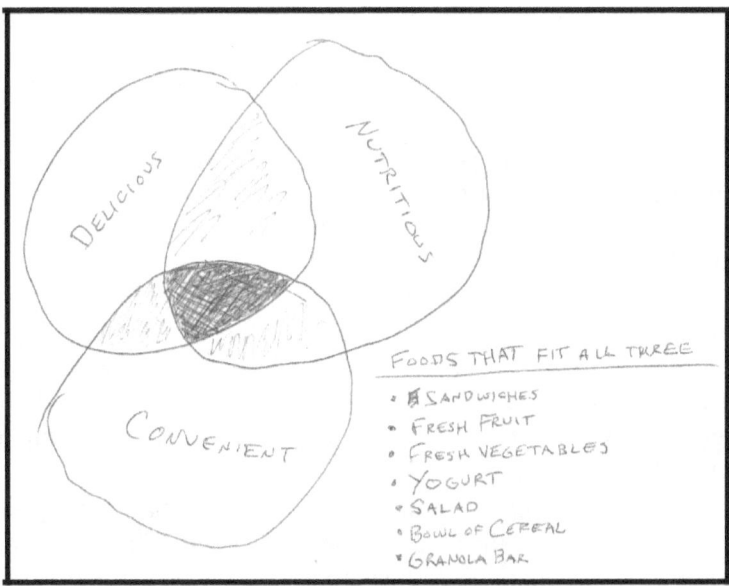

Scan of the actual Venn diagram drawn in the hospital

The items we came up with that fit snugly in the middle of all three were things like fresh fruit and vegetables, yogurt, salad with fat-free dressing, a bowl of cereal, granola bars, and homemade sandwiches. I was really impressed and was proud of the boys for their creativity and excitement to get involved with this. Then Nathan told me that we forgot to put chicken nuggets and cat food in the middle. So I went back to the beginning and explained it to him again…

Gross Motor Skills

A lot of factors impacted my gross motor dysfunction. I was put into hypothermia to slow my body functions down enough to help compensate for the severe hypoxia (oxygen deprivation) I was suffering from. Then I had to be given a paralytic while in hypothermia to prevent shaking and tremors. I had been unconscious and in roughly the same position for nearly a month. I had heavy tubes attached to my legs. I had undergone several surgeries. I woke up from sedation unable to feel or move my right arm from the shoulder down, or my right leg from the hip down. My stomach and groin were similarly numb and insensate.

After getting over my initial shock and momentary fear that the paralysis was permanent, the doctors explained that it wasn't uncommon for some numbness to occur. Unfortunately, there was no way to predict how long it may take for sensation to return – and they could never guarantee that it would. Within a week, however, I started recovering "high level" sensation. If someone put their hand on my leg and pushed down, I could feel the pressure and direction, but not necessarily where on my leg they were pushing.

Eventually more and more sensation returned, and I was able to get my arms and legs to move when I told them to. Most of the time. It was arduous at first. Rehab consisted of trying to flex my leg or lift my arm off the bed. Then repeat it over and over again. Once I was able to get my legs under me (literally) I had to contend with my balance.

My sense of balance was way off. The extreme weight loss moved my center of gravity just enough that I felt like I was constantly tipping forward. My blood pressure was low enough that the slightest changes to elevation would make me light-headed. Even turning my head from side to side too quickly would make me dizzy.

After regaining some initial strength, my rehab routine was to stand up at the edge of the bed, then sit back down. I had a four-post walker with me at the time, and had to hold onto the handrails for support. At first, I had to pull myself up with my arms, but eventually my legs started to help out.

The first time the therapist took the walker away and tried to get me to stand unassisted, I just sat there. I didn't even budge. She tried to encourage me and told me to at least give it a try. "I am trying!" I told her, but my legs were still too weak to actually lift my entire body mass. It took days and days of practice before I was able to stand using just my legs – but I often still needed a cane or other support for balance until I was completely up and had a second to adjust.

Fine Motor Skills

Working on my fine motor skills was also a challenge. I was really surprised at the variety of different tasks and tools the therapists used. There are so many different muscle groups and systems that work together to perform very specific movements. There were different exercises to work on my pinching strength than used for something that seemed very similar to me, such as using a fork or pair of scissors.

I don't remember all of the tests or resources used, but a few come to mind. There were all manner of standard, expected tools and exercises. A few were unexpected. Sometimes the more unusual or unconventional the tool, the better I responded.

One of the first tools I got was a small plastic tub filled with bright pink putty. That putty was my constant companion for weeks. There were actually more than a dozen different exercises to perform with the putty alone. I could work on my grip by squeezing the entire mass. Work on my pinching by tearing pieces off. Work on flexion by stretching and pushing it with my fingers. I could work on sensitivity and touch by rolling a small ball of putty between my thumb and fingers. The putty barely left my hands. My hands were sore, but they started working again, so the pain was worth it. I'm not sure I would have been able to hold a pen without the putty.

Which leads to my second fine motor exercise. Writing. I wrote a lot. I hand wrote more than a dozen pages of notes during my incarceration (er, hospitalization). I hand wrote thank you cards to as many people as I could. I wrote down TV channel listings. Nurses' names. Whatever I could think of. My handwriting was atrocious at first, but improved rapidly.

It wasn't easy, though. Writing started out as a battle of wills. My mind versus my hand. I could see the pen in my fingers, and the pen nib on the paper. But I couldn't feel the pen at all. And without that tactile context, I had a hard time knowing how much pressure or strength to apply to write. My first fumbling attempts were big herky-jerky motions. I'm sure it looked like Frankenstein's monster flailing about.

Once I got into the one-on-one sessions at the Courage Center, I was able to suggest a few things that the therapists added to their repertoire. One of the OTs asked me if I liked boardgames. (Well, how much time do you have to discuss it?) She wanted to play checkers to help me with fine motor control, pinching, and balance. One of my symptoms at the time was that the further my hands got from my body, the more they trembled and shook.

I noticed they had Blokus on the shelf, and suggested that instead. I told her we could kill two birds with one stone – practice all the motor skills she had mentioned, plus give the brain a bit of a workout. Blokus is a bright, colorful puzzle game that's almost like multiplayer Tetris. You place tiles of different shapes and sizes on the board, trying to use as many of your tiles as possible before running out of room. It was awesome. And it kept me from having to play checkers.

The most interesting test I performed required hand-eye coordination, sensory acuity, and fast reflexes. I had to stand in front of a large board that had three concentric circles printed on it, the outermost circle not quite as wide as my arm span. Each circle had eight glowing buttons spread equidistant around its circumference – so there were 24 buttons total. In the center of the middle circle was a small LCD display. I had no idea what to expect.

Then one of the buttons lit up and a number flashed on the LCD display. I had to hit one of the buttons while saying the number that appeared on the screen. I had less than three seconds before a different button and a new number would appear. The first time I tried it I think I was only able to successfully touch 10 or 12 buttons before time ran out. Once I figured out what the test was all about, I wanted to do it again. Probably about 20 button matches the second time. Again! I can do better! This was like a motion sensor video game like Xbox Kinect or the Wii... I got this.

I think my best was close to 40 matches. I don't know how many tries it took me to do that well. But I do know that exercise completely, utterly wiped me out. It was probably the best sweat I had built up during rehab, including time on the treadmill or bike. Unfortunately, I wasn't prepared for the consequences. My arms throbbed the rest of the day, and the following morning, I could barely use them they were so stiff and sore. Worse off, my right shoulder had swollen up like a balloon – starting a whole new series of tests to see if I had a pinched nerve, dislocation, or other problem.

Building Endurance

Looking back, it seems absurd that a trip to the bathroom could exhaust somebody. Or sitting up in bed and dangling your legs over the side could make you tired. But after your body hasn't done anything but lay there, motionless, for a month? Even those small tasks become workouts.

It was hard not to get frustrated by how easily I got fatigued. And not just tired like you need a nap. I felt like every ounce of energy had been drained out of me. It was utter exhaustion all the way down to the bones. Even my hair was exhausted.

My endurance improved slowly while I was still at Regency, but at least I was able to see progress. Once I got to the Courage Center, things really took off. With the level of autonomy I had at the Courage Center, I had to become more self-sufficient. And with so much time on my hands and nothing to do, practicing my balance, or standing up or eventually walking sounded much more appealing than sitting around and watching television all day. I could do that once I got back home.

The first few days, I needed a wheelchair to get around. I quickly graduated to supervised walking. When I first started walking again, I had to wear a special harness and be accompanied by a spotter to help me keep my balance. Eventually I outgrew the need for the harness, then the supervision.

Each day it seemed like my endurance doubled. I could push myself in my wheelchair twice as far. Eventually I could walk twice as far. Or go up twice as many stairs. Or perform the fine motor exercises for twice as long. I would still end up completely wiped out by the end of the day, but at least there was a growing sense of accomplishment and measurable progress.

That continued to improve as I transitioned to outpatient Cardiac Rehab at the Heart Center at Methodist Hospital. I had rehab sessions scheduled three or four times each week. On Mondays and Tuesdays, there were classes to attend after my sessions. There were a variety of topics, from making diet changes to dealing with the stressors of a heart condition. Even though I had already gone through all of these classes the previous September, I attended most of them again. Hopefully more of it would sink in the second time around.

During my cardiac rehab sessions, I was hooked up to a portable EKG monitor so they could keep track of my heart rate and pulse during the entire workout. One of the physical therapists would stop by to take my blood pressure before, during, and after my cardio routine.

My first session, I couldn't even walk for five minutes on a treadmill. By the time I had completed my rehab goals, I was on an elliptical for 30 minutes per session. While that worked up a good sweat and got me breathing hard, my heart rate, pulse and blood pressure kept looking good. I also started to notice that my recovery time after each workout kept improving, as well. Finally I was discharged from the program to work out from home on our elliptical, which is great.

I also remember one interesting encounter during cardiac rehab. On my second or third day, one of the physical therapists was looking over my file and she said, "Oh, you're *that* guy! I heard about you." I had no idea what she was talking about. Then I found out that her husband was the EMT who sat in the back of the ambulance with me when I was first transferred from the Methodist ER to the University of Minnesota. Small world, huh?

Emotional & Spiritual Recovery
This is still an ongoing process. And probably will be so for a long, long time. What I went through was almost inconceivable. It was a death-defying, life-changing experience. I am literally and figuratively a different person now than I was before my heart attack. I have had to struggle with processing countless feelings, anxieties, memories, and emotions, both real and imagined.

The entire event also had me question some of my beliefs and faith. While I never doubted the existence of God (quite the opposite, this entire experience reinvigorated my belief in God), I was suddenly unsure of how I fit into His plans, or of what really is important spiritually. The pomp and circumstance of a structured Mass at a church suddenly seemed... well, a little bit silly.

Over the span of weeks, I spent more time in prayer, reflection and meditation in my hospital bed than I ever had in all my years of attending church. Possibly combined. And I felt closer to God and His presence in my life than I ever had in church. I didn't use formal, structured prayers. I didn't follow ritual or tradition. I just spoke, straight from my heart and soul. Every day I was in the hospital, I would say a prayer as soon as I woke up, thanking God for seeing me through the night and letting me open my eyes to a new day. If I felt myself drifting off to sleep, I would say a quick prayer just to be able to wake up. And I got the feeling that God was more receptive than ever before.

A priest stopped by on two different Sundays to offer Eucharist and a blessing. Over the course of my stay at the ICU and Regency, I was surprised by how infrequently spiritual assistance was available. There were only a handful of visits overall, even after requesting time with whoever was making the rounds – chaplain, pastor, priest, deacon, rabbi, minister... at the time I didn't care what religion. Just that they believed in something bigger than all of us. That's what I needed.

It may sound strange, but I spent very little time wondering *"why me?"* That seems to be something everyone assumes I spent a lot of sleepless nights struggling over. There were only a few times during my hospitalization that "why me" even entered my mind – one of them being when a man in an adjacent room had died. And that was more *"why him and not me this time?"* As odd as it sounds, instead of feeling relieved and comforted by surviving such an ordeal, there was a period of time when it made me wonder what was wrong with me.

If the afterlife is a place of peace and tranquility, if Heaven is where the soul yearns to be, why was I denied not once, but twice? Do I have some unfulfilled purpose? Is there a meaning to my life that I don't understand? Am I not good enough to get in? Or is it simpler than that – maybe God didn't want my children to grow up without a father. And despite my soul's best efforts to leave this corporeal vessel and reach Heaven, God said, *"Not yet. You've still got work to do."*

I've always struggled with my purpose in life, why we're here in the first place, what we're meant to do… I've been plagued by deep philosophical and spiritual questions that there are no answers for. These constant questions lie at the root of my insomnia and depression. It was hard enough just to understand the physical part of my ordeal. Trying to fathom the mental and spiritual impact left me stupefied at times.

I've seen therapists and counselors, psychiatrists and psychologists, but those practices never seem to pay off. Part of it is that these people have no vested interest in me. Once I leave their office, they might take down a few notes. But then I'm out of their lives and thoughts until the next session. Another part of the disconnection is that they simply cannot relate. They have not gone through what I have. I've had plenty of people to talk to who haven't shared that experience – I don't need another one of those. But I have yet to find someone who has.

Sure, there are thousands of people who have survived heart attacks. And there are support groups out there for those people. I happen to be 20 to 40 years younger than most of them. I felt that despite the shared experience of a heart attack, there just wasn't enough of a common bond or common ground to achieve the sort of connection I think I'm still looking for.

What I really need is a near-death experience support group, much like the way alcoholics or gambling addicts need a support group to help them cope. But it's such a niche need, I haven't been able to find anything of the sort. And rediscovering life has kept me too distracted and busy to start one of my own.

15. For the Record

After returning home, I requested medical records from each of the facilities I had stayed in during my recovery. Paperwork comprises the bulk of my records, as one would expect. What I wasn't expecting were CDs with the video of all my angio procedures. They feature the dozens of pictures taken during my angioplasties, and string them together into mini-movies covering different parts of the procedures. It's all right there, in stunning black and white.

There are also lots of ultrasound images of my heart from the numerous echocardiograms conducted. The good news is that my heart isn't pregnant. The not so good news is being able to see exactly where the damage to my heart occurred. Both of my ventricles are in bad shape, as is the ventricular septum or wall of heart muscle between them. And since heart muscle is one of the few tissues our body doesn't replenish, they won't get any better on their own.

Now that The Event feels so far removed, it is not quite as terrifying to read my medical records. But it is rather unsettling. Ok, I take that back. Every once in a while it is completely terrifying after Googling a term or when realization finally hits on what a certain paragraph is referring to. This chapter looks at my medical records and some of the interesting information found within.

And boy, do hospitals keep detailed notes. I have more paperwork than a ream of paper – easily 500 pages – and that's not even all of it. That's only the documentation from my 24 hours in the Methodist Hospital ER and my time in the University of Minnesota Cardiac ICU. There are also dozens of folders, brochures, pamphlets, and pieces of literature regarding the procedures I had, the medical equipment used, and even explaining some of the techniques employed during my surgeries. Then last but not least are those CDs.

I Have to Admit Something…

Or in this case, admit some*one*. Me. One of the interesting parts of going through my medical records is reading through the admission and summary notes as I move from one facility or specialist team to the next. These notes contain a small summary of everything that's happened up to that point, and my current condition at the time. Seeing some of the same notes and comments over and over again is a little bit… strange.

May 23, 2011. Exam On Admission

[verbatim from Methodist Hospital ER records]

Temperature 98.5, respiratory rate 22, pulse 66, blood pressure 140/93, O2 sats are 100% on room air. Exam was unremarkable, with regular rate and rhythm without murmur, lungs clear to auscultation, abdomen with bowl sounds present, soft, nontender, nondistended. Labs revealed a normal white count of 7.9, hemoglobin 14.6, platelets 207, normal BMP, calcium, troponin less than 0.1, CK 71, with EKG with a normal sinus rhythm with ventricular rate 64 beats per minutes, with chest x-ray negative for any acute disease. INR on admission was also 1.0, PTT 26.2. Hospital Course: Patient was admitted and monitored. He remained asymptomatic and pain free overnight.

[end of notes]

This is from the day I first felt the symptoms and drove myself to the emergency room. It's alarming how mundane and perfectly normal everything is, just hours before my heart attack took place.

May 24, 2011. Hospital Course

These notes were taken while still at Methodist Hospital, after moving from the ER to the floor.

[verbatim from Methodist Hospital records]

Patient was admitted and monitored. He remained asymptomatic and pain free overnight. His troponins were negative x2. He was sent for exercise stress echocardiogram, which was read as negative. He returned to the room pain free, however, within 10-20 minutes after returning to the room, he had nausea, vomiting, diaphoresis, with 8/10 chest pain on left, and then had a syncopal episode with heart rate of 80s, soon went into ventricular fibrillation without a pulse. Code Blue was called.

[end notes]

May 24, 2011. Assessment on Admission

These notes are from after my heart attack, once I was transferred from Methodist to the University of Minnesota.

[verbatim from Fairview Hospital Record after transfer]

A 38 year old male with a significant medical history of NSTEMI s/p stent to3VV (Ramus, OM, and LAD) who is transferred from Methodist for ACS and cardiogenic shock. After the pt was transferred to CCU,

SWAN GANZ was inserted. After discussion with CVTS [Cardio, Vascular, Thoracic Surgeons], the decision was made to put him on ECMO and the patient will be transferred to SICU. His family was updated, per the note from Methodist. Later on, his wife was present and plan was updated.

[end of notes]

Interesting side note. Under the section titled "Chief Complaint" on that day's record, they wrote down: *Unable to obtain a history from the patient due to critical condition.* At first I thought it was kind of rude to expect me to provide a medical history while I was, you know, in a coma. Sorry for the inconvenience, guys.

But then I realized it was referring to the chief complaint that brought me to the ICU in the first place. Duh.

May 25, 2011. Cardiology Evaluation

[verbatim from hospital notes by Dr. Cindy Martin]
Pt arrived to 4D [my room] hypoxic and requiring massive amounts of vasoactive drugs thru peripheral IV.... X-ray confirmed ETT position and reveals severe pulmonary edema. Pt pupils were reactive and showed some spontaneous movement despite apparent prolonged hypoxia therefore decision was made to proceed with ECMO and cooling protocol. Pts' wife was updated on pt's condition and grave prognosis.

[end of notes]

I shudder to think what would have happened if my pupils hadn't contracted. Would they have not initiated the ECMO? Would that have been the end? Instead of "grave prognosis," what would the notes have read? What would they have told Trish? Was my fate really determined by how my pupils reacted?

It turns out, that just might have been the case. In one of my follow-up visits, Dr. Martin confirmed that I was in dire straits. The doctors had to consider the possibility that I was already brain dead after such an extended period of profound hypoxia. At that point, the doctors weren't certain I would ever wake up from sedation – or if I did wake up, how severe the brain damage would have been.

If the eyes truly are the windows to the soul, I'm glad my soul closed the windows rather than trying to sneak out through them.

May 26, 2011. Cardiac Evaluation Consultation

[verbatim from hospital notes by Dr. Ranjit John]

REASON FOR CONSULTATION: ...evaluation of options for cardiogenic shock with respiratory failure.

HISTORY OF PRESENT ILLNESS: The patient is a 38-year-old gentleman who was transferred from Methodist for acute coronary syndrome with cardiogenic shock... [myocardial infarction recap]... The patient was also hypotensive and intubated, started on dopamine and epinephrine with intra-aortic balloon pump. He was then transferred to the University of Minnesota Medical Center in severe cardiogenic shock with respiratory failure.

REVIEW OF SYSTEMS: Unable to do as patient is intubated, sedated and paralyzed.

[end notes]

June 20, 2011. Pulmonary Consultation Assessment

[Verbatim from Fairview Hospital Record]

[summary of myocardial infarction event]... The patient has a complicated hospital course, including approximately 1 week of ECMO for hypoxia and cardiogenic shock as well as placement and subsequent removal of an intraaortic balloon pump. The patient also underwent induced hypothermia and has since been rewarmed. He was extubated approximately 3-4 days ago and was doing fairly well; however acutely this morning around 3 a.m. he was reintubated for acute hypoxic respiratory failure. At that time, his sats were in the mid 50s and he was tachypneic [breathing rapidly]. The chest x-ray showed significant worsening of peripheral upper lobe opacities as well as opacities in the left base.

[end of notes]

June 24, 2011. Progress Notes and Assessment

[Verbatim from Dr. Trehan's Regency Hospital Notes]

[the patient] was able to tolerate a speaking valve and was very happy about that. The patient also stood up today and did some therapy, and this was a good day for him. He is recovering from cardiac arrest and sudden cardiac death in the setting of coronary artery disease with cardiogenic shock, at this point stable.

[end of notes]

Just the casual reference to "recovering from death" caught me completely off guard. Some people have trouble recovering from the common cold.

July 7, 2011. Subjective Patient Assessment

[Verbatim from Dr. Trehan's Regency Hospital Notes]
The patient is resting in a chair. He feels comfortable and is at peace...
In terms of coronary artery disease, he is doing well post cardiac arrest. He does not have any recall of what happened but at this point is stable.
[end of notes]

September 28, 2011. Admission Diagnosis

[verbatim from Fairview Hospital Record for ICD surgery]
Mr. Jason Little is a 38-year-old gentleman with a history of myocardial infarction twice, complicated with cardiogenic shock in May 2011. The patient also had an in-hospital cardiac arrest which was resuscitated. Repeated echo afterward showed that the patient has an ejection fraction of 25-30%. He underwent ICD implantation today and was admitted overnight for observation.
[end of notes]

December 31, 2011 History of Present Illness

These notes are from my unexpected ER trip on New Year's Eve 2011. These were the most intense symptoms I felt since my discharge. There were several symptoms experienced in a short period of time, so I erred on the side of caution by going to the hospital. Not how Trish or I expected to spend our New Year's Eve together.

[verbatim from Methodist Hospital Record]
This is a 38-year-old male with a complex past medical history. He had a non-ST elevation MI in September 2010 and had a drug-eluting stent placed to OM1. He returned to the Cath Lab 2 days later for a staged procedure on his LAD and first diagonal, both of which had drug-eluting stents placed. He had repeat angiography in October 2010 showing patent stents. In May 2011, he had a recurrence of chest pain similar to his MI. He ruled out by serial troponins and underwent stress echocardiogram, which was negative. However, when he returns to the hospital floor, he suffered a ventricular fibrillation arrest.
[end of notes]

So What Actually Happened to Me?

Piecing that puzzle together is harder than it looks. Notes are scattered throughout my records, as different doctors and specialists focused on different aspects of my health. But all the information is in there. I just had to be patient enough to sift through everything. Patience is not one of my virtues, but I gave it my best shot.

After taking the cardio stress test on Tuesday afternoon, May 24th, I returned to my room to rest. I was pain free for approximately 10-20 minutes. Suddenly I started vomiting, and was sweating profusely. I complained of intense chest pain and cold and hot flashes. Then, all the medical terms cloud the details.

I experienced a syncopal episode with a heart rate of 80s, and went into ventricular fibrillation with no determinable pulse. The doctors called a Code Blue. I was given epinephrine, bicarbonate, and amiodarone. I was shocked with a defibrillator repeatedly, trying to resuscitate me. I would come out of v fib briefly before experiencing more tachyarrhythmia.

Next, I was taken to the cath lab for an emergency angioplasty. That's when this took place: "angiogram demonstrated prox LAD thrombotic occlusion which was stented with DES."

It sounds like after the catheritization, my condition was what would be termed severe cardiogenic shock. They inserted an intra-aortal balloon pump to help my heart function. Finally, when I was stable enough for transport, I was taken to the University of Minnesota.

At the U of M, one of the first courses of action was to insert a SWAN GANZ line and surgically hook me up to the ECMO machine – and yes, the attachment and removal of the device is classified as a surgery in my records. That is some sophisticated piece of equipment.

I remember none of this. I do remember going to take the stress test, but nothing about the test and nothing afterward. Everything was learned secondhand from my doctors, family, and medical records.

Now in Layman's Terms
Short story: I had a heart attack. A big one. But you knew that already.

OK, Now the Longer Layman's Version
The syncopal episode essentially marks the point where I fainted, or at least lost all consciousness. The ventricular fibrillation (or "v fib") is when my heart beat and pulse started to go all wonky, likely causing inconsistent delivery of oxygen to my brain, hence the fainting.

Then my pulse stopped altogether, and the doctors called a Code Blue – the medical term used to denote a patient needs immediate resuscitation. So that's when the defibrillators came into play.

When I was shocked, I'd come out of the v fib, but fall into tachyarrhythmia – or in other words, my heart would start beating again, but abnormally fast and at an uneven or abnormal pace. My heart kept stuttering and starting.

Eventually I was taken to the cath lab for an angioplasty. During an angioplasty, the doctor inserts a small catheter into one of the larger blood vessels along the groin, and snakes it all the way up through the network of blood vessels to the heart. They found a blood clot that had completely closed off my left anterior descending (LAD) coronary vessel. That's the one that supplies most of the horsepower to the heart.

When you look up LAD or LAD Occlusion online, damage to that vessel paints a grim picture. The Wikipedia entry has a section in the LAD article titled *The Widowmaker*. That made me squirm before I even started reading the entry. It states:

> *Because the LAD provides much of the bloodflow for the left ventricle, which in turn provides much of the propulsive force for ejecting oxygenated blood to systemic circulation via the Aorta - blockage of this artery is particularly associated with mortality. In the medical community ischemic heart attacks associated with this blood vessel are colloquially called 'the widowmaker'.*

So I had one of those widowmakers. Thankfully it didn't live up to its reputation. But it still messed me up pretty good. The trauma led to cardiogenic shock, which ain't good. That's the point at which my heart can't support the other organs in my body, because it's too damaged to perform its job. That is one reason I had the intra-aortal balloon pump inserted – it had to help pump the blood for me, because my own heart could not do it alone.

During this time, I was also classified as "profoundly hypoxic." In simpler terms, my body was oxygen deprived. Important organs and systems weren't getting the oxygen they needed to function properly and were starting to shut down. I think this is why some of the medical records indicate I also suffered a stroke – because for a dangerously long amount of time, my brain wasn't getting oxygen.

Upon arrival at the U of M, they inserted a SWAN GANZ line into my neck. This is another type of catheter with ultra-fine filaments. The line is threaded directly into the chambers of the heart, allowing doctors to gather cardiac information from inside the heart itself. I assume this gave doctors the most accurate readings and data, rather than other methods that would measure activity around or outside the heart.

This is also when "emergent ECMO was initiated." I was surgically hooked up to the machine that provided extracorporeal membrane oxygenation. In other words, my blood had to be cycled through a big machine stationed next to my bed. The machine artificially infused my blood supply with oxygen since my heart and lungs weren't up to the task.

In reading about ECMOs, it was interesting to discover that the machine seems to be used far more for respiratory failure than cardiac failure. Since I had both, I guess I was a good candidate. One disturbing statistic to read was that the survival rate for a patient requiring ECMO is generally only around 50-70%. Then I found out that survival rate is artificially on the high side, because it takes into account ECMO's highly successful results when used with neonatal emergency care. Trish was told the success rate was closer to 10% for an adult to come through without any long-term debilitation, if they even survived.

At this point, pretty much all my body functions relied on electricity and things that beeped. My body was running out of oxygen, and ultimately could not sustain itself.

Everything Else Falls Apart

I've always had a flair for the dramatic. Apparently I don't even need to be conscious. Oh no, my body wanted more attention. My medical records from the University of Minnesota detail eight separate diagnoses requiring a hospital course of action to manage during my initial care. Because four or five additional problems just wouldn't be enough.

1. **Cardiogenic shock from acute STEMI.** This has already been covered. It's basically the heart attack part. The medical records go into a lot more detail, with metrics, medications, percentages, and a ton of acronyms.

2. **Acute hypoxic respiratory failure.** Considered secondary to my pulmonary problems. When I read that I thought, *"I guess if the heart stops working, the rest doesn't really matter."* Hypoxia is the state in which your body isn't getting the oxygen it needs to function properly. My records indicate I was in severe hypoxia for a "prolonged duration (at least 2 hours)."

3. **ARF.** No, not barking like a dog. I had Acute Renal Failure – my kidneys stopped working. Dialysis was required to treat this, off and on through the first few weeks.

4. **Shock liver / elevation of transaminases.** Bzzt – my liver was fried. Elevated levels of the wrong enzymes, impairing my liver's ability to function properly and to keep me from looking like a Simpsons' character.

5. **Anemia.** Low blood iron. Iron is vital for healthy blood oxygenation. This was treated with transfusions, to introduce some friendlier blood to my system. I had a "less than robust response" after the first two units were transfused, so several more blood transfusions were required during my critical recovery time.

6. **There is no sixth thing.** My records actually skip #6 in the big list of problems, and move right on to #7… Dunno if it was a typo, or if later on they decided my blue painted thumbnails and one dark painted toenail weren't indications of something more severe (like cyanosis) than just my quirky personality.

7. **Infectious disease.** Oooh, I had *coagulase negative staph*. I also developed pneumonia and VRE – *Vanccomycin Resistant Enterococci* – over my hospitalization. VRE is a potentially serious respiratory infection that hung around for a while. And I also contracted *candida albicans*. Which is funny, because I don't remember ever visiting Candida.

8. **Deconditioning.** I would need significant occupational and physical therapy after a prolonged hospital stay and subsequent muscular atrophy. I lost a lot of weight in a very short amount of time, and a significant amount of that was muscle mass.

But Wait, There's More

Everything in that previous list only looks at the *first week* while I was in the Cardiac Intensive Care Unit at the University of Minnesota. I wasn't even awake for most of this. Heck, the doctors were probably able to cross most of these items off their To Do list before I even had a chance to know I had suffered from them.

But more happened to me than just my heart attack. I was still in and out of sedation for another week or so. Some symptoms or conditions couldn't even be diagnosed until after I had regained consciousness and more tests could be performed. Here are some of the other repercussions from The Event.

1. **Post-Traumatic Stress Disorder**. Yes, it's the same kind of anxiety disorder soldiers returning from the Iraqi War suffer from. My body underwent shock, but so did my mind. Given the nature and intensity of my delusions, part of me thinks I could have suffered PTSD from those alone. But once the ramifications of the entire medical event were made clear, and I understood just how unlikely my survival was, the PTSD settled in. I go into the mental impact a bit more in other sections of the book, so no more detail is needed here.

2. **Sepsis**. When the lay term for something is "blood poisoning" you know it's not going to be good. Sepsis is a "potentially deadly medical condition" [Wikipedia]. In simplest terms, it is the severe illness that results once the bloodstream has been completely overwhelmed by bacterial infection.

    The condition of being septic may reference SIRS – Systemic Inflammatory Response Syndrome. I had the misfortune of having severe sepsis. This is characterized by three things happening to your body at the same time, when just one would suffice: the inflammatory response, a serious infection, and organ dysfunction. I've always been an over-achiever.

    The only thing I could remember about sepsis off the top of my head was that a good friend of mine in St. Louis was a doctor who had once studied the effects of sepsis on lab rats, and the results weren't pretty.

3. **Distal, Ulnar and Radial Neuropathy**. I've recovered from a great deal of the nerve damage suffered during the ordeal. I've regained sensation in my extremities, and along my leg, stomach, and groin. But some significant damage remains. Enough time has passed now that after several EMG tests and consults with neurologists, it's uncertain if these symptoms will ever fully recover on their own.

    My left arm suffers from Ulnar Neuropathy, which runs along my elbow, down into my pinky and ring fingers. It feels like my funny bone is almost constantly being triggered, causing intense pain along my left arm. My right arm suffers from Radial Neuropathy, reducing my range of motion, and overall hand and pinching strength, as it manifests most strongly in my thumb and forefinger.

These symptoms are also associated with carpal tunnel syndrome. So fortunately, we're able to try remedies that address CTS, and are trying to manage the pain with medication. Unfortunately, 90% of my job requires fine motor skill use, specifically working on a keyboard, so performing my job necessitates aggravating my symptoms.

4. **Memory Loss**. What was that? Oh yeah, I forgot a lot of things. Not uncommon for someone coming out of a coma or suffering hypoxia for so long. I had a combination of both long-term and short-term memory loss. I had trouble recognizing people I apparently knew well.

    At first, a lot of this was attributed to the heavy duty sedatives I was given during those first few weeks. But even after those wore off, we could tell there were some gaps and missing information here and there. I had trouble recalling details from grade school, high school, and college. I wouldn't recognize photos of people I apparently knew. And I would become easily confused, forget what I was doing, or get lost trying to follow directions along a route I had taken many times before. Granted, while some of this happened even before the heart attack, it became far more common afterward.

5. **Kidney Stones**. Other than being extremely painful, these were mundane. Not mutated, or radioactive, or infected or anything special like that. Thankfully this is no longer an ongoing issue.

Putting Things into Perspective
Sorry, I just can't. Not everything, at least, and not all at once. There are some times I think I've got it, then a certain detail or memory comes up and my mind is blown all over again. There are a number of things about my hospitalization and heart attack that I only learned through the research for this chapter – I don't remember it being told to me by someone else. That cast everything in a whole new light.

Between the previous heart attack, my history of heart disease and hypertension, the LAD infarction, the need for an ECMO, the heart necrosis and weakening, the prolonged hypoxia, the pulmonary edema, the respiratory failure, the pneumonia, the infectious diseases, the renal failure, the liver shock, and the sepsis... I had better odds to hit the lottery jackpot or be struck by lightning

I mean, come on. Who lives through all that? Apparently I do. Now I understand why so many doctors and nurses called my recovery "unbelievable," "remarkable" or "miraculous." Because after reviewing my medical history and reading over everything that happened to me, *nobody expected me to survive.*

I can certainly see why. The odds were stacked against me, big time. Looking at it from that perspective, my survival just doesn't make any sense. It can't, unless you're willing to acknowledge non-medical factors. Yes, the medical technology, knowledge, and staff were absolutely essential, but with everything that happened, it seems absurd to discount some sort of spiritual assistance or divine intervention. In my case, I firmly believe it was a combination of intercession-by-love (i.e. Trish calling me back, and me listening) and God's presence.

So thanks, God – I owe You one (million).

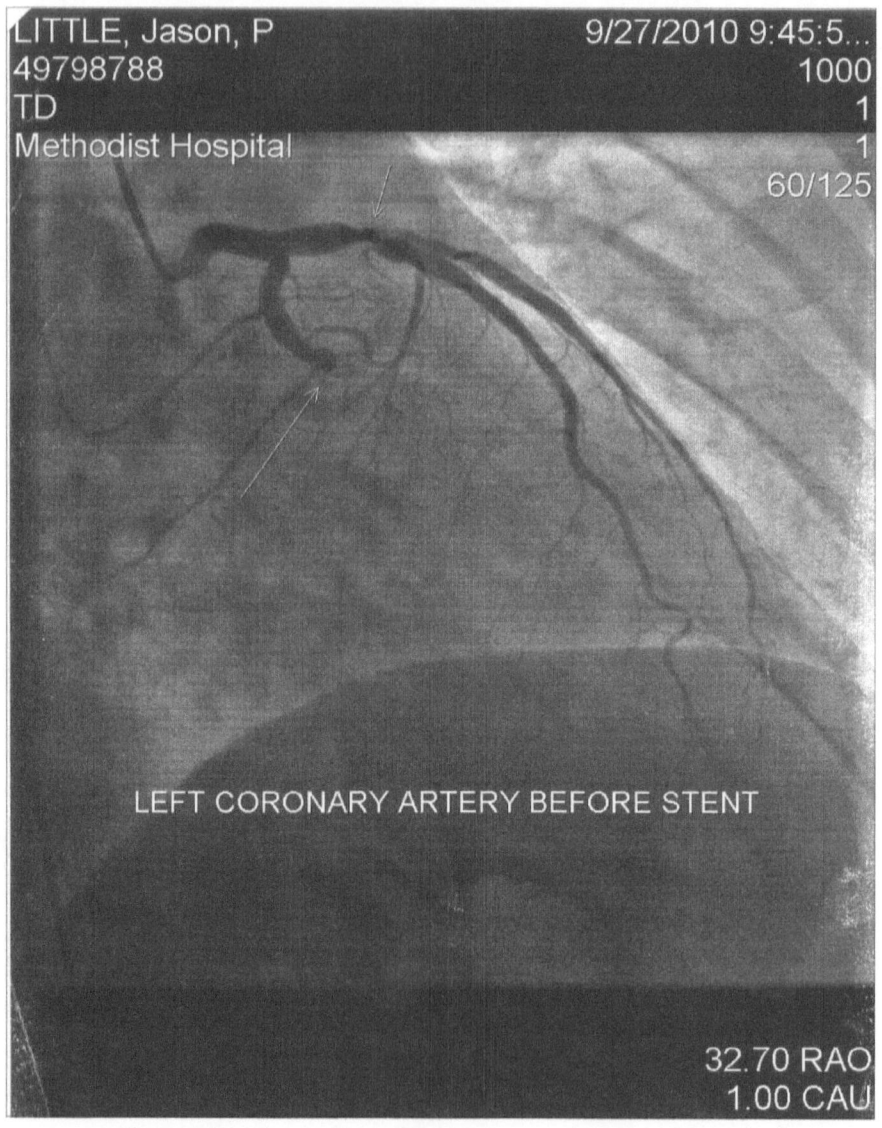

Here is a cath lab image of one of my occluded vessels after my first heart attack in September 2010. The vessel just... stops, pinched off by the occlusion. This road is temporarily closed for repairs. Find an alternate route. My blood had nowhere to go. The appearance and thickness of the vessel between these two photos is as different as night and day. Or life and death.

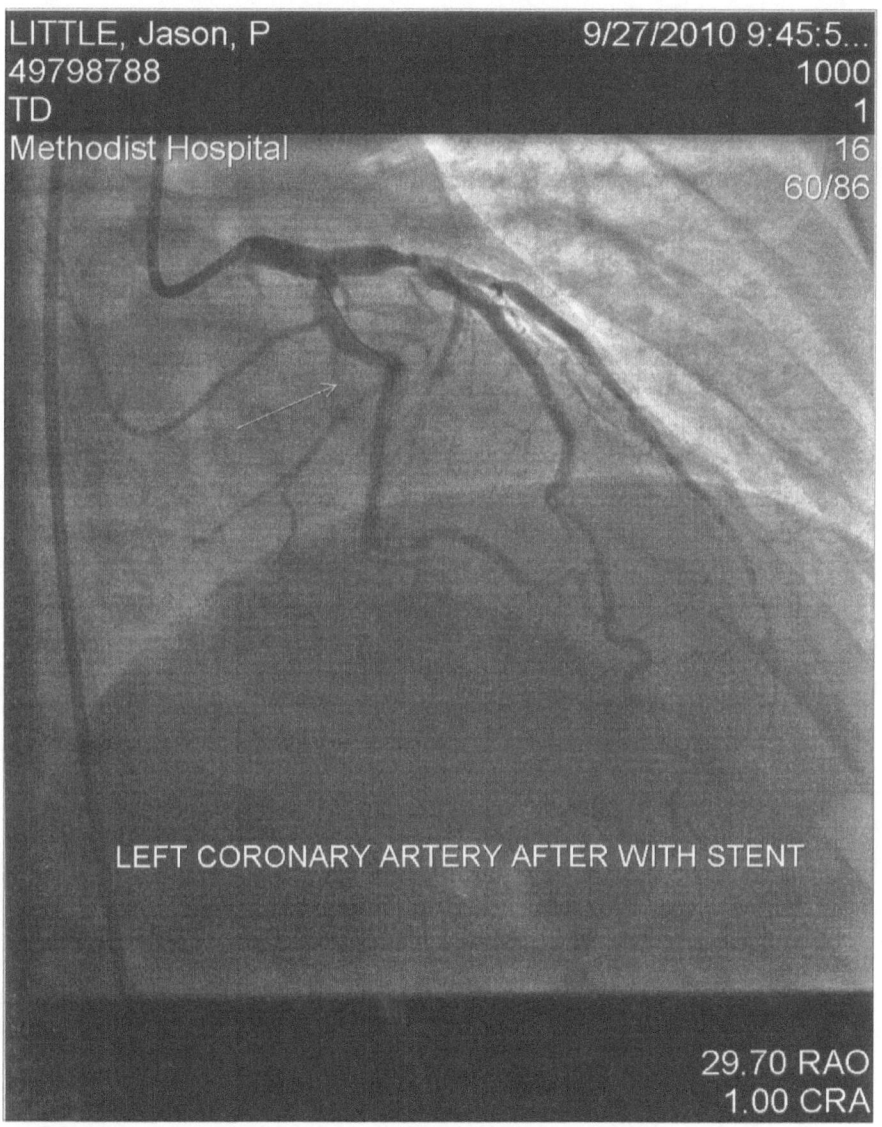

And here is a cath lab image after the angioplasty procedure in September 2010 to clear the clot and insert a drug-eluting stent. The stent is essentially an incredibly small spring or coil that expands inside the vessel to help keep it open. The difference is just amazing. By the end of it all, I had to have four stents inserted to open up my blood vessels.

16. The Idiot Box

You can't lie prone for weeks on end without something to distract yourself from time to time. And what better way to aimlessly waste time than the idiot box. I watched a lot of television. Way too much. I'll keep this chapter brief. Essentially, it's just a list of the TV shows I remember watching, with a few notes sprinkled in here and there.

- **Travel channel**. I remember watching a lot of shows that had to do with families relocating overseas for new jobs. Like moving from New York to India, and the sort of culture shock they encountered.
- **Numbers**. Unfortunately, this was one of the few choices at 3 AM. I ended up watching a lot of episodes, because I ended up being awake at 3 AM a lot. I remember math making me sleepy back in high school. It's too bad it still didn't have that effect.
- **Criminal Minds**. I couldn't stand this show, but again, it was one of the few shows on in the wee hours. If you made a drinking game out of every time someone said "unsub" when referring to their suspect, you'd be drunk before the episode even started. Lame.
- **NCIS**. I ended up addicted to this show. There were two episodes on back-to-back each afternoon, and it wasn't long before I went from having seen zero episodes, to being caught up through the first three seasons and jonesing for more. Even after returning home, I kept watching this.
- **Scrubs**. Good show. Great cast. But not quite as funny when you're watching a comedy about hijinx in a hospital when you're actually in a hospital.
- **Food Network**. I constantly tortured myself by watching all the foods I couldn't eat. *Chopped* and *Diners, Drive Ins & Dives* are particularly good at inducing stomach rumbling. I still watch them both frequently, and still can't eat anything shown on either show.
- **New Looney Toons**. A crime to all that was once fun and entertaining about the original Looney Toons. This was a disgrace.
- **The NuWave Oven**. One of my favorite infomercials at the time. Probably watched it at least a dozen times. And I do want to make food faster, healthier, and more delicious. Sign me up!
- **Sensa**. I saw this infomercial for the weight loss supplement quite a bit. Shake these little crystals on your food before eating, and watch the pounds drop off.

- **The Baby Bullet**. It's like our big, modular Bullet blender system with all the cool attachments – except smaller, and now only good for blending soft things for homemade baby food.
- **Ninja Blender System**. Another great infomercial for the ultimate blender. It chops. And dices. And slices. And makes bread dough. And ice cream. And crushes ice. And makes restaurant quality frozen drinks. Is there anything the Ninja system can't do? I actually really want one of these.
- **FIFA Soccer Tournament**. I think it was FIFA time. Anyway, lots and lots of soccer matches were televised – that much I do remember. I wouldn't say I'm a big soccer fan, but I watched a lot of matches, and actually got a little caught up in it. One of my doctors was British and another from India, and they were both big soccer fans. They would stop by my room to check in on the scores and see how their favorite team was doing.
- **Baseball**. Thank goodness the hospital got Fox Sports Midwest. I was able to watch a few dozen Minnesota Twins baseball games. If I had my heart attack just a few months earlier, in that lonely gap between the Super Bowl and start of pre-season baseball, my recovery would have gone a lot worse.
- **Border Wars**. This show fascinated me. Both the extent to which drug runners and other criminals would risk their lives to get into the US, and the guts of the US marshals and law enforcement who had to protect the border with inadequate resources, face constant danger, and perform largely unappreciated duties.
- **Early AM News**. I watched a lot of super early news shows. The ones that start before the sun comes up. I'd watch and see the traffic cam reports and slowly witness the morning commute build up then slow back down. I'd be tormented by the weather reports of how wonderful and gorgeous it was outside. And then there'd be a report on a kitten that rode skateboards or something.
- **Charmed**. Sucker for this. Paranormal. Magic. Hot chicks. 5 AM. This became one of my "stories" or regular soap opera fixes.
- **Angel**. Beloved as he is for *Buffy the Vampire Slayer* and *Firefly*, Joss Whedon doesn't get nearly enough credit for how awesome *Angel* is, and how well it still holds up.
- **Burn Notice**. It's *The Bourne Identity*, the series. What's not to like? Plus it has Bruce Campbell and his chin, so it's even better.
- **Royal Pains**. My number one guilty pleasure at the time, and my biggest soap opera crush. I didn't realize how addicted to the show I had become until Trish came to visit me after I had just watched every single episode during the Season 1 marathon and was trying to

explain this love triangle, and the awkward feeling between the brothers because of this other thing, and how these other two are struggling to share their true feelings, and… then looked at her, saying, "and that's probably more than you ever needed to hear about the show."

- **Unnecessary Roughness**. Another sleeper hit from USA network. Psychotherapy and sports. Good combination, entertaining show.
- **The Closer**. Great cast. The quality of the show just didn't quite match up to the quality of the actors in it. But it was on a lot, and I didn't mind watching it.
- **Suits.** After Royal Pains, this was probably my favorite show. USA was just cranking out a bunch of compelling shows with interesting characters. Fake lawyer playing real lawyer, doing better than all the real lawyers… and trying to avoid being caught as a fraud.
- **Covert Affairs**. Soft-core spy series with the beautiful Piper Perabo
- **White Collar**. This is almost like *Catch Me if You Can*, the series. Con man now using his skills to help the FBI crack cases – but still has secrets and schemes churning in the background.
- **Psych**. This started in the hospital, but became one of my "churn" shows. These were the shows where entire seasons were available on Netflix, so I would watch an entire season in just a few days. James Roday and Dule Hull are phenomenal as a fake psychic and his obsessive-compulsive, squeaky clean friend.
- **Leverage**. A crew of Robin Hood-esque con artists who set up sting operations and grift their way to exact revenge and serve their personal brand of justice. Damn I love this show.
- **Fallen Skies**. It's V, only with more special effects and robots. And Moon Bloodgood. She's so dreamy.
- **Franklin & Bash**. Young, crude, sophomoric lawyers who just happen to be too damn good to get rid of. More entertaining than I thought it would be.

In the news

Aside from all the entertainment, I distinctly remember two big news stories unfolding during my recovery. Every station seemed to have some sort of coverage, and the news scrolling along the bottom of the television kept us constantly updated. Whether we wanted it or not.

Since one of the stories involved a little white girl disappearing, Nancy Grace was on television yelling about it every day. Funny how I can't remember her covering a story about any missing Hispanic or African American children.

Rupert Murdoch Newspaper Scandal

I don't remember specific details leading up to the sudden collapse and dissolution of one of his newspapers in England… Just something about investigative journalism / sensationalism run amok by tapping in to private phone lines or answering machines. Invading privacy to gather dirt and dig up clues, while somehow offering false hope to families thinking there was activity on phone records or logs from their loved ones, not from reporters digging around.

Somehow this behavior from a glorified *National Enquirer* surprised a lot of people. This story was constantly on, as more details about their practices were made public. But I don't know why so many people thought *this* was across the line, given the lengths paparazzi have gone for years to get a story.

The Casey Anthony Trial

You could not get away from this. Whether you wanted to or not, you were immersed in the Casey Anthony trial through television osmosis. I swear every channel had some sort of Casey Anthony coverage. I wouldn't be surprised if it was part of ESPN's bottom scroll along with all the baseball scores. It seemed to be the only thing people in the hospital were talking about for a while.

The nurses were all abuzz with the case. Every shift change, the nurses seemed to flock together to share anything new they had heard, or talk about how Casey Anthony would surely be convicted. Or how they can't believe what so-and-so said on the stand. Or how the mom could not possibly have intervened or known what was going on. Or what a <bleep> Casey Anthony was for going out and partying and boozing it up rather than being a responsible parent.

All the nurses were absolutely convinced she'd be found guilty. Despite the prosecution's case relying heavily on circumstantial evidence, it did seem pretty damning. I wasn't obsessed about the verdict, but I definitely thought she was going to be found guilty.

I remember hearing gasps and shouts of "No!" from the hallway when they announced the not guilty verdict. The tension, frustration, and disbelief were palpable. The only other time I can remember anything close to that sense of "everyone is watching" across the country was with the OJ Simpson trial.

17. The Discouragement Center

It's not pleasant, but it warrants mentioning my time at the Courage Center. The Courage Center came highly recommended by the staff of Regency. It is an acute therapy center that helps people recover from severe trauma. While I had already recovered some of my faculties and ability, I was still a good candidate, so I was transferred there on Wednesday, July 13th.

The updates to CaringBridge related only one side of what occurred during my stay. What follows is the official grievance I filed with the Courage Center in August 2011. While my overall experience with the Courage Center was extremely negative, it's important that I acknowledge a few very, very bright spots indeed.

So let me share the good before I delve into the bad.

Washington and Charla were two of the best nurses/aides I had over the course of my entire recovery. They were responsive, always cheerful, and encouraging. Washington always had a smile on his face, and Charla had a great, mischievous laugh.

Lori and Sue were my two occupational therapists, and they were spectacular. One helped me with my gross motor functions, and the other focused on my fine motor skills. They were always pleasant and attentive, and encouraged me just the right way that I pushed myself to keep trying and practicing rather than giving up when I'd struggle.

I met so many people in much more dire straits than I was in that it helped me gain perspective. I wasn't just coming out of a nine-month coma like the first resident I met. I hadn't been hit by a train and broken more than half the bones in my body like the other Jay I met – my age, with two daughters the same ages as our two boys. His family was wonderful and there a lot – his mother-in-law and I talked quite a bit over my time there, and it felt good to talk to someone else who had at least partial insight into what I was going through.

Oh, and the grounds were lush, green, and well maintained. It was a beautiful, state-of-the-art facility.

Unfortunately, those are the only positives I can come up with. So on with the grievance.

I'm writing to file grievance for the numerous issues that occurred during my stay at the Courage Center from Wednesday, July 13th through Saturday, July 23rd. Many of these concerns were brought up with several staff members along the way, as well as with Mary Martin, my stay supervisor. There were so many issues and failings that I felt it important to send this letter in hope that future residents will receive better care from the Courage Center.

I must say this is especially frustrating given the excellent reputation the Courage Center had among staff at different facilities, and it was based on the recommendation of others that the Courage Center was chosen for my next stage of care – had these people been aware of the numerous issues, I doubt they would have been so willing to offer glowing recommendations.

Rooming Conditions

Unfortunately, many issues stemmed from my roommate. I understand that having single occupant rooms is impractical, but a lot could be done to improve the living conditions given the fact there are two people in each room.

A good example of this is when evening medications or blood draws are scheduled. Rather than only turn on the lights over the bed of the person receiving treatment, all the room lights are turned on – so you would always be woken up when your roommate received treatment, and vice versa.

Regarding my roommate, while I had done my best to be respectful and sensitive to his condition and needs, the situation grew increasingly frustrating. It had reached the point where I wasn't just uncomfortable, but I did not feel safe in my own room.

1) My roommate required a balance alarm to monitor when he would attempt to move or transfer in ways he wasn't cleared for. The balance alarm was triggered frequently day and night. It is a high-pitched alarm which was impossible to sleep through. Even more frustrating were the times it would beep for 20-30 seconds or more before any staff responded to the alarm. In fact, one evening the alarm went off four times within a five minute period and never had a staff member check to see what was the matter.

2) On several occasions, my roommate pressed the call button for assistance and received none. When the reception person buzzed in to see what he wanted, he was unable to communicate or speak to alert them of what exactly is needed. This usually resulted in one of two situations – he would buzz again (sometimes with no response), or he would grow impatient and attempt to move himself (setting off his alarm).

    I had to intervene on his behalf and buzz to the front desk to ask for assistance several times, once when he was close to falling, once when his alarm kept sounding and no one came to check on him, and once when he started wheeling himself out of the room with no clothes.

3) The bathroom situation was disgusting. I had to clean urine from the floor and toilet seat several times before being able to use the bathroom. After showers, I had to move the towels, wash cloths, and soap out of the way since they were left strewn about. Personal hygiene items were left out. After he shaved, hair and whiskers were left all over the sink -- no one cleaned it up.

4) His visitors were smokers. The pungent smell of cigarette smoke clung to them, often so strong I needed to leave the room when they visited. I don't know for sure, but I suspect they were smoking on the Courage Center grounds (which is supposed to be a smoke-free campus), since they've entered with particularly strong cigarette stench after going outside for a walk. The ash-tray stench lingered long after they left, making it difficult to rest, relax, or try to sleep.

5) The most dangerous situation occurred on Sunday the 17th. Somehow my roommate had access to a lighter and a pack of cigarettes. While watching TV in the adjacent bed, I heard a scratching noise from his side of the room then start smelling smoke. Concerned, I checked and saw that he lit up and started smoking right there in our room! I alerted the nurse, and several people had to come in to resolve the situation and eventually get the cigarette and from him.

    Not only did this pose a serious health risk, it stunk up the room considerably. I told the staff I did not feel safe in my room and wanted to have one of us moved as soon as possible. Thankfully he was moved out of my room within 36 hours after bringing the issue up with Mary Martin, every nurse, and attendant I could.

Finally, while being discharged on Saturday, August 23rd, I was amazed to see one of the staff place a Bic lighter on the information desk at the entryway. My former roommate, now directly across from the information desk, saw the lighter, grabbed it, and headed back to his room to smoke. I had to alert the staff again about this hazardous behavior.

There were a number of other concerns, as well.

Needs Not Being Met

I felt some of my needs simply weren't being tended to. One of the most important things I mentioned several times before admission and during welcome to the Courage Center was the ability to talk to a psychiatrist. The event I suffered which eventually lead me to the Courage Center was traumatic and frightening, but there was never any mention of when or how the mental side of my recovery would be handled, even though I had asked several times.

The scheduling was hectic and unreliable, and most staff could only respond with a "wait and see" answer. I finally did meet with a therapist earlier in the week after pressing the issue more, and we had a brief introductory session. Afterward, she mentioned we should schedule another session and definitely see me before the end of the week. Not surprisingly, the week went by without another session.

Diet & Cafeteria Management

There were a number of issues with the overall management of meal planning and food services.

I was surprised at the number of times I would see food services staff without wearing hair nets or gloves while preparing and serving food -- even though boxes of gloves are on the table right next to the service area. After first observing this, I started to keep tally in my notebook on how many times I'd see staff not wearing gloves. From my brief stay, I stopped keeping track after the eighth time, resigned to accept this wasn't an important issue for the staff.

The food services staff was also ill-equipped to handle special diet needs. Even though each patient had their own dietary card on file with the staff, which they would pull and confirm who I was during meals, it made no impact. Even after reviewing my special diet restrictions (low sodium, cardiac diet) I would have staff offer me soup (never a good sodium option), a cheeseburger, one poured cheese sauce all over one entree while I was trying to tell her that certainly doesn't fall into my diet -- I had to have her re-make the meal.

One evening I told the staff member rounding people up for dinner that I was having dinner with my wife rather than at the cafeteria. They still set aside a meal for me, and the woman who delivered it reprimanded me for missing dinner. She brought me an omelet with cheese, pancakes covered in butter and syrup, and sausage links – nothing that fit within my diet restrictions.

For having to meet with a nutritionist during my stay, it is hard to believe there is no dietary information available for any of the food being served. No one could provide information on calories, fat, or sodium levels for any of the individual items or overall meals. A menu for the week was posted in one of the common rooms, but no foods on the printed menu were flagged or identified as being low sodium, low calorie, or had any other information that would be useful to a patient with specific dietary needs and restrictions.

Medication Management & Nursing

The management of my medication was erratic, especially timing -- at times asking me to take my morning medication before breakfast, with nothing in my stomach. At night, I had requested several times to have all my evening medications delivered at the same time. Most often it was split into two or three separate trips, even though the delivery schedule easily allowed them to be combined.

Sleep was a constant issue, so the timely delivery of my Ambien was very important. That was one of the medications usually delivered much later than the others, and much later than I had requested. One evening, I was even woken up by the nurse to take my sleeping pills!

At one point during my stay, they ran out of medication. One day I was told they had run out of Heparin, so I wouldn't be getting my next few shots for that -- potentially missing two doses of one of my most important treatments. That was quite frightening, until they finally got more Heparin and were able to resume my treatment schedule.

The quality and professionalism of the nursing staff varied widely – too widely in some cases. One of the nurses attending me constantly forgot to bring certain pills, water to take my pills with, and transposed numbers in my blood pressure; she'd say it was 118 / 71 (for example) but then later say it was 111 / 78. When she couldn't find a pen to record it, she took my pen from my desk, and ripped out a small part of my notebook to write down the information – tearing off a piece of paper I was working on! When I noticed her writing down the wrong numbers, I had to have her take my blood pressure again.

Dr. Weurmser & Professional Service

During orientation, I was told I would meet with my assigned doctor, Dr. Weurmser, within a week. Possibly more quickly than usual since my stay was expected to be very short. For a stay of only ten days, I did not meet with my doctor until the end of my eighth day. And that appointment did not go well.

I was Dr. Weurmser's last appointment of the day, and it felt like the appointment was very rushed, with her constantly watching the clock. Aside from a brief overview of my condition, she ordered an X-ray for my right shoulder, to be taken the following morning (Friday). I still had outstanding questions and wanted to know the next steps to be taken, and she said she needed to evaluate the X-ray results first. It felt so rushed, she "wrapped up" my appointment while wheeling me out of the office toward the elevators.

Sure enough, the X-ray technician arrived early Friday morning to take the X-rays of my shoulder, so they would be ready for Dr. Weurmser to review as soon as possible Friday and she could propose a next course of treatment. I did not hear back on that Friday about the X-ray. I contacted them Monday (August 25) and was assured they'd have the results by the end of the day. They never called back on Monday.

On Tuesday (August 26), my wife and I each left several messages, and were assured by the person who took one of our calls that she would look into it, and we would hear back by the end of the day. They never called back on Tuesday.

On Wednesday (August 27), I called again and expressed my growing frustration over the mismanagement of my X-ray. I was told by the person I spoke with (Val) that she would see to it personally, even if it meant contacting Dr. Weurmser on her day off, and that I would have the results by the end of the day. They never called back on Wednesday.

Thursday (August 28), I called and was unable to get to a live person. I left a very detailed message with Val stating I no longer had confidence in Dr. Weurmser or the Courage Center's ability to manage my health. I requested all my medical information and X-ray results be forwarded to my Primary Care Physician, so I could be treated by a doctor I trusted. They never called back on Thursday.

Friday (August 29), my wife had to drive me back up to the Courage Center so I could meet someone face to face and finally resolve this. It had gone beyond inconvenient to unacceptable. Only after committing nearly an hour of drive time and an almost 30 minute wait at their offices, were we finally able to get the results we had been asking for over a week's time and numerous phone calls.

Amazingly, when we reviewed the reports, the X-ray results were actually available by Friday, 22nd at 11:30 AM. So they had information available to pass along to me for a week, and despite calling and requesting it nearly a dozen times, no one ever called us back with an update.

Therapy Sessions

While schedules were erratic and constantly shifting, it was difficult as a patient to build any rapport with the physical therapists. For my first five physical therapy sessions, I saw a different PT each time. So they each had to review all the previous notes and had no context for my current condition or limits. This made it very difficult to set goals and measure progress.

Information sharing was also an issue. The therapists clearly did not have the most current information on my status. For example, my 7 a.m. Wednesday occupational therapy session was to help take me through my morning routine – a shower, brushing my teeth, and so on. However, I had already been doing all of these things on my own since the previous Friday.

The Discouragement Center

As you can imagine, these issues made my stay extremely uncomfortable and disappointing, especially after reading and hearing about the Courage Center's reputation. I can only imagine it would be worse for the residents who cannot advocate for themselves, or for patients less assertive than me and insecure in bringing up issues.

There is only a very slim silver lining to my stormy stay at the Courage Center. I can say that the occupational therapists (Lori and Sue) were excellent, and several of the attendants and nurses were responsive, attentive, and cheerful (Eniola, Betty, Colleen, Washington, Hope, and Charla). Unfortunately, the rest of my experience was very frustrating, discouraging, and left me feeling insecure and concerned about my health, with no clear direction for therapy or treatment.

Jason Little
[contact information]

I sent this letter to Mary, my care coordinator at the Courage Center, but I also sent copies to the Minnesota State Ombudsman for Healthcare Advocacy and to a reporter at the Star-Tribune who had written articles about patient advocacy and conditions at clinics in the area.

To their credit, someone from the Courage Center called me. Mary had forwarded my grievance to the new director for the Courage Center, who I believe had only been in the position for a few weeks. I can't remember her name, but she called me to personally apologize and say she was shocked that my stay at the facility she had just assumed control over went so poorly.

She told me not to worry, she would take care of everything and help close out and resolve my stay. Apparently I had a very different definition of "take care of everything" than the Courage Center did. A few months later, we got a bill from the Courage Center for their services not covered by insurance.

I was furious. They had the *gall* to bill me after the way I was treated and my case was mishandled? Someone named Nancy even called to collect and remind us of our outstanding balance. Trish was going to just pay it and be done, but I wouldn't let it go.

We played phone tag for a week, and I finally gave up. After a week of silence, Nancy started calling again. Finally, I was able to return her call when she was there instead of leaving a message. I asked her if there were any special comments or notes attached to my record, because there certainly should be – and if not, she needs to contact my care coordinator and ask for a copy of the grievance I filed.

All I got from her was "*blah blah*, not my department, *blah blah*, I can't do anything about that, *blah blah*, please pay your bill." I told her the director of the Courage Center had personally called me and told me she would take care of everything. So talk to her and confirm everything and stop billing us.

Not good enough. Finally, I told her that I know any sort of medical billing system has discretionary leeway for special cases – special payment plans for a low-income family, or even writing off any entire balance as a donation or otherwise retroadjudicating the charges. She said that was beyond her decision making level. Well then, she better talk to someone who can make those decisions.

Apparently she did. The following week I received perhaps the best phone call ever. She called and quickly told me that they decided to simply cancel the billing and take care of it internally, and then she immediately hung up.

It's hard to express just how smug and victorious I felt.

But it was a lot.

18. Introspection

As you have no doubt discovered by now, an awful lot happened between September 2010 and September 2011. In the span of one year, my body was subjected to severe trauma, and our family subjected to severe anxiety, fear and uncertainty. As more time passes, some of the anxiety and fear diminishes, or gets replaced with more mundane, less life-threatening uncertainties.

But it's not something I think I will ever truly be "over." Like losing a limb or losing a loved one, the Event was a keystone. The Event became a linchpin to the rest of our lives, whether we liked it or not. It was a big, fat domino that toppled over, crushing everything directly beneath it, but in the process knocking over dozens of other dominoes racing down their own separate paths to… who knows where.

Some of those domino paths have already reached their conclusion, and with thankfully positive outcomes. The pneumonia, renal failure, sepsis, paralysis – each of those spiraled outward from that initial domino. Some of the other paths continue to click-clack along, knocking over more and more dominoes along the way.

The lingering anxieties and fears hang overhead like storm clouds, big, dark and ominous. And there I stand underneath them, without an umbrella. Nobody can predict what the storm will do. It may lash out with lightning and torrential rain. It might be nothing but a light drizzle. Or it may even move along, posing no threat at all. But it's hard to stand beneath that blackening sky, the sound of thunder drumming in your ears, and not fear the worst.

This chapter gets a little trippy and philosophical.

My Very Own Storm

The analogy to a storm brewing feels more and more appropriate as time passes. Forces are gathering. Energy inexorably pulled toward some unknown purpose, with unpredictable but potentially devastating results. But you can't live your entire life just waiting and waiting for something that may never happen. Then again, you can't afford to be caught unprepared, just in case something *does* happen.

The series of health scares and my ongoing health issues are at the eye of the storm. It has affected or has the potential to affect so many different parts of my own life, as well as the lives of my family and loved ones. Some of those effects are easy to understand. They are the effects that make sense. Just like rain makes sense coming from a storm, some are downright obvious.

Food for Thought

My diet has changed dramatically. Now I'm eating more fiber and veggies, lower protein portions, more fluids, and less sodium. I'm supposed to have 2,000 mg or less of sodium per day. But old eating habits are hard to change. A fast food meal, some take out Chinese or barbecue could easily have twice that in one "serving." I feel like I have to be careful with each and every bite I take.

Since my heart attack, I do the majority of the grocery shopping and cooking. I'm incredibly label conscious, and have made sure we have more fresh fruit and vegetables on hand, for meals and for snacking. I've found a number of great recipes that allow me to make and enjoy some of our favorite dishes with little or no added sodium. But even with these added steps and a greater overall awareness of healthy eating and good habits, food is an enormous complication.

I struggled with depression even before my heart attack. And the struggle has only gotten worse. Bouts of depression have routinely triggered episodes of binge-eating. Now the ramifications of that are more dangerous than ever. Suddenly eating smaller portions or less sodium doesn't matter if you end up eating two or three extra meals in a day or burn through an entire bag of snacks in one sitting.

I can ruin an entire week's worth of healthy eating and exercise with one evening of binge-eating. And I stress even more about my fluctuating weight. I weigh myself almost daily, plotting huge swings of weight gains and losses, cringing each time I step on the scale, wondering what it will be today...

Dark Days Indeed

Depression is serious stuff. Not just being sad, but utterly depressed. Feeling hopeless or distraught. Losing interest in your favorite activities. Feeling like a burden to those around you. Distancing oneself from family and friends. Even with medication and counseling, these have been part of my life for as long as I can remember.

And it's easy to see how everything we've been through can contribute to depression. The heart attack and health concerns have certainly affected me psychologically, and episodes of depression are unfortunately even more regular than before.

The worst part of one of these black moods is a voice in the back of my mind screaming at me that what I'm doing is wrong, pleading with me not to eat something I shouldn't, or trying to convince me I shouldn't be dwelling on something. But the thundering dark storm clouds drown out that voice, no matter how loudly it tries to shout.

Not every day is doom and gloom. In fact, there have been a number of incredible days. Once released from the hospital, I've easily had more good days than bad days. There have been moments of peace, where I can count my numerous blessings and appreciate everything we have. There is an indescribable sense of wonder and amazement at what happened, and a deep appreciation for medicine, prayer, and miracles.

But even all that can't shed enough light to brighten my mood or clear my mind when depression hits.

When it Rains, it Pours

Not all of the side effects are so obvious. My storm unleashed a lot more than just rain. Looking back over the last year, and taking time to reflect on everything for this book, a few things snapped into focus for me. There are some thoughts and issues I find myself going back to over and over again. I'll lay there and ponder these questions for hours whenever I struggle to fall asleep.

How Much Longer?

This is perhaps the most frustrating question that keeps coming to mind. The nagging question that pops up each time a symptom manifests. As a rational, practical person, I realize that everybody dies. Some people die younger than others. Some people die unexpectedly. And I obviously know that I'm going to die someday. But when? That's the part that drives me crazy.

My health issues and the trauma from my heart attack have affected my life expectancy and longevity. But since nobody knows how long that was in the first place, how can you measure or predict how it's been affected? Coronary disease and heart failure are deadly, but they are not a terminal form of cancer. No doctor can tell me I only have a few years or months to live. Only that the longer I live, the greater my risk of not living any longer than that.

The good news is that while writing this book, I was told I am no longer at high enough risk to warrant a heart transplant. While it's pretty dinged up and severely damaged, my heart works well enough to keep me going for the foreseeable future. With proper exercise, diet, and other lifestyle changes, I might get another 100,000 miles or so out of it.

What is a Soul?

This is the biggest spiritual question raised by The Event. It's at the heart of what makes me "me." What is a soul, and how has it been affected by everything? Has my soul been damaged, just like my body?

It's another unanswerable question. But that doesn't keep me from asking it over and over again. Early in 2012, I scheduled an appointment with the priest from our local parish to sit down and talk about how the event affected me spiritually. And the conversation quickly turned to a discussion about what our beliefs and convictions are about our souls. Unfortunately, he was unable to provide answers. I'm not even sure what I was expecting or hoping for. Despite the church's dogma asserting the existence and importance of our souls, there isn't much else to go on. Not without stepping out of traditional dogma and wandering off into new age spiritualism.

Having nowhere else to turn, I looked inward. Since my heart attack, I have spent an inordinate amount time in contemplation and prayer. I realized that the only place I would find an answer was deep down, by sifting through my own thoughts and feelings. Now I understand why they call it soul-searching.

And what did I discover? Nothing earth-shattering or prophetic. Unexpectedly, I found something comforting and reassuring; I know I have a soul. How do I know this? I can feel it – deep down in my soul. It's one of those things that doesn't make sense, proving something's existence in a self-referencing loop.

Perhaps the best way to describe it is to say that my prayer and reflection kept following a certain circuit or path. It was like a Mobius strip – a line of thought that slightly twists on itself. But if you continue to follow it, you'll eventually end up back where you started, despite the twists and turns along the way.

The more I contemplated my soul, an image started to form in my mind. The analogy my mind came up with is a hermit crab. Our mortal body is the shell, and our soul is the tiny crab within. Over time, our shell may crack or chip, or maybe the crab has matured and simply outgrown its current shell. When it's time for the crab to move on, it must discard its older shell.

Who Am I?

The hermit crab analogy really stuck with me. It gave me a line of thought to explore and contemplate. With each heart attack, my shell cracked. The first was just a small crack, and despite the inconvenience, my soul didn't feel the need to find a new shell.

But the second heart attack nearly smashed the shell to smithereens. And I think for a time, while I was comatose, my soul was ready to find a new shell. Mine was so badly damaged, it wasn't sure if it could survive in that shell anymore. But for whatever reason, be it Divine Intervention, stubbornness, or medical marvel, my soul decided to stick around and make do with this damaged shell.

But this analogy for my soul didn't stop there. During my recovery, there was another part about the hermit crab analogy that started to resonate with me – the crab grows over time. It matures. And eventually, *it moves on*. This is when my interest in the nature of my soul shifted focus to a much wider view. One that encompassed the concept of past lives.

While I consider myself rational and intelligent, I also believe in God, miracles and spirituality, which some may find contradictory. So how much of a stretch is it to believe in past lives, within a spiritual framework? I have faith that it's not a stretch at all.

I started wondering if some of my delusions weren't complete fabrications. Instead, perhaps they were memories or experiences my soul had previously lived through. And its current damaged shell was doing the best it could to make sense of it all.

Two delusions in particular took me down this path. The first is of my daughter, Rainbow Joy Little, the girl that never was, which I chronicled in *Chapter 9: Jedi Mind Tricks*. Was I misremembering that experience, and my mind simply plugged Trish and I into the roles of the grieving parents – had that actually happened to my soul's previous shell? The feelings and images are too intense and realistic otherwise. Despite leaving its former shell, my soul still grieves for its daughter.

The second delusion is based on when I fought in the American Civil War. This isn't one I discussed earlier in the book; it's even more detailed. I'm no history junkie, but I know a little bit about the Civil War. One of my good friends in St. Louis was a Civil War buff, and I've watched a show or two on the topic on the History channel. I'm no expert by any means. So how in the heck did I know all of this..?

Marching into Hell

I was there, fighting for the Confederacy. I was part of the Virginia Corps on the third day of the battle of Gettysburg. It was the third of July. It was hot and muggy. There were rumors that General Robert E. Lee and Lieutenant General James Longstreet had been arguing about the upcoming day of battle. Eventually word came down through our division leader, Major General George Pickett.

The camp was chaotic, as we rushed to ready ourselves. We were about to assault Cemetery Ridge and Major General George Meade's Union forces. The Virginia Corps were the spearhead of the attack. Our flanking maneuvers the night before hadn't been successful, and General Lee could tell our morale was flagging. So we were chosen to carry out his plan to bloody the Union's nose and help push our way northward. If successful, we could end the war today. Or that's what we were told.

But this was madness. I may have been infantry, but I wasn't stupid. This was suicide. We were being thrown into hell to try and buy time for Lee's other maneuvers. This must have been what Longstreet had been arguing about – surely he could see this plan was doomed to fail before we even started.

It wasn't up to me, though. So we were drawn up into lines. Then we charged. On foot, across what seemed like miles of open ground. Right into the teeth of an entrenched Union position. And every single one of us knew we were going to die that day. But we charged anyway. We had to. It was our job, our responsibility.

The cannonade and artillery support we were told would hammer away on the Union position never came. That gave the Union cannons free rein to wreak havoc. The Union artillery batteries and infantry opened fire. The noise was so intense I was deafened. There was only a sharp ringing in my ears. All around me was smoke, blood, and gore. I saw dozens of our soldiers literally torn apart by the canister and gunfire.

So many of my friends and fellow soldiers littered the ground we literally had to march over and through them. Sometimes we had to step on their fallen bodies to advance. Then it felt like my shoulder exploded and I fell. I had been hit. The remains of my mangled arm were lying a few feet from me. Blood and darkness clouded my vision, as I watched more of my comrades shouting and rushing forward. More of them were hit, fell, and screamed. Thankfully I blacked out.

The horror I witnessed was so real. The intense pain of being shot was excruciating. The bleak hopelessness I felt when I realized I was being sent to my death and would never see my wife again – it is all so intense and real that a part of me simply cannot believe it is all a lie. That part believes this memory must have been one of the times when my corporeal shell was destroyed and my soul had to move on.

What Am I?

All this introspection and soul searching raised far more questions than it answered. The growing sense of wonder at what my soul really was raised another issue. If my soul is an eternal entity traveling from body to body while it matures, what does that make the rest of me? Is my life as I know it simply playing host for a soul as it travels through its current iteration?

The shell analogy comes back to mind. An empty shell. I think I was empty for a while, my broken body just a broken shell hooked up to machines. But once my shell started to recover, my soul wandered back and made itself at home once again. Maybe if my soul had been mature enough it would have moved on to heaven, or if my body had been dead any longer, perhaps it would have found a new host elsewhere.

But I think my soul heard Trish when she whispered, "come back to me." Something brought me back, something convinced my soul to turn around and… re-integrate with the rest of me. And I can't think of a more powerful incentive than love.

Given my tastes for fantasy and science fiction, this train of thought soon took a detour. It stopped being a question of *who* I was, and slowly became a question of *what* I was.

For argument's sake, if I believed the soul and the body were discrete, separate entities that usually mesh together so well they are inseparable and unaware of their individual uniqueness – then what happens when one part suddenly becomes aware of the other? Is what was lying there in the ICU just an empty physical vessel while my soul thought about moving on? And once my soul snapped back into place, what does that make me? A golem? Zombie? Homonculus? Some sort of biologic automaton? Are we symbiotic? Am I the host – or the parasite?

Thinking Too Much

My head starts spinning every time I go down this path. It's a long, spiraling journey, and it's easy to get lost. Sometimes, I just think too damn much. I can't seem to shut my brain off when I need to. This is one of the reasons I have so much difficulty sleeping. I keep *thinking*.

It's like some heroic quest that is impossible to complete. Or maybe the twelve tasks of Hercules' atonement. Except my impossible task is the same as millions of other people over the course of humanity – the yearning to understand the meaning of life. I don't think I'm any closer now than I was before. But if nothing else, this entire event did reveal one undeniable truth to me. That truth is I am a paradox, a living contradiction. *I am the same person I was before, yet I am no longer the same person I was before.* It's so unerringly true it's mind-boggling.

Stepping back to try and gain some perspective, it seems like I spend so much time pondering the meaning of life, I forget to actually live it.

19. By the Numbers

Reading my medical records was both fascinating and horrifying. To see that so much happened, over such a relatively short period of time, but involved so many different people – it's amazing. One of the things that really stuck out from all of the paperwork, conversations and copious notes were the numbers. Numbers everywhere. Numbers for everything. Here is a list of a few of the more interesting numbers regarding my cardiac event, overall hospitalization, and subsequent recovery.

185
Estimated number of heparin shots administered over the course of my stay. I had received so many injections my entire midsection was bruised. It was a wide purple belt of bruising that wrapped all the way from hip to hip across my stomach. Toward the end, the nurses simply couldn't find an un-bruised section for injections.

14
Number of times I needed to be shocked after the initial heart attack before my heart resumed a normal enough rhythm to be stabilized. This number amazed me. Pop culture and movies show the doctors zapping a patient two or three times before giving up – they gave it their best shot and have to move on. I wish I knew how long a period of time this encompassed.

428.20
The hospital tracking code for Systolic Heart Failure. It's never encouraging to see "heart" and "failure" together. It appears a lot in my medical records.

140/93
My blood pressure when I arrived at the ER on May 23rd 2011.

102/76
My blood pressure on release from the Courage Center in August 2011.

19
Number of different items on my current medication list.

< 0.1

My troponin levels (an enzyme used to measure heart activity after cardiac events) at the baseline, 90 minute, and six hour mark on each of my five ER visits since my first event in September 2010. This implies a completely normal test, with no indicator of a cardiac event. This includes the ER stays during both of my heart attacks.

50

Number of pounds I lost during the first four weeks of my hospitalization. I was comatose for the vast majority of this time. It's not a weight loss program I would recommend to anyone. I was so gaunt, I barely recognized my own face in the mirror.

39

Number of different tests listed on my medical record on May 24th alone, just at the Methodist Emergency Room, before my transfer to the University of Minnesota

4

Total number of showers I received from the time of my heart attack until my transfer to the Courage Center. I recommend better hygiene than that if you can manage it.

28

Average number of pills I take in a day, +/- 2 depending on symptoms

1

Millimeters of ST elevation in V1 through V3, associated with T-wave inversions consistent with ventricular aneurysm. I have no idea what that means. But between my Q-waves, T-waves and upsloping ST elevation, reading my medical records often felt like reading a Scientology indoctrination pamphlet.

0

Number of times I was seen by my doctor my first week at the Courage Center. I only ended up seeing her once, right before my discharge.

33%

My ejection fraction in August 2011, once the doctor felt it had normalized enough for a reliable measurement. This was approximately ten weeks after my heart attack. Ejection fraction is essentially a measure of the heart's overall efficiency, more specifically the output of the ventricles. A healthy adult male will usually have an ejection fraction around 50-55%. Anything lower than 40% is the threshold at which my cardiologist explores other options, such as a LifeVest, ICD, LVAD, or heart transplant. I ended up with an ICD.

23%

My ejection fraction during my hospitalization approximately nine months after my heart attack, on 1/1/2012. Not a comforting trend.

4

Number of medicine-eluting stents I currently have placed in various coronary vessels. Between the stents and the ICD, I'm well on my way to becoming a cyborg.

34

Number of weeks between my two heart attacks.

3

Number of healthcare professionals who came into my room on June 5th and just started running tests or tried to draw blood without introducing themselves or showing me their ID. I didn't want anyone near me that I didn't know, or wasn't completely sure was from the hospital.

7

The percentage of my daily recommended sodium intake that should come from any one source. For reference, a single slice of wheat bread has 8% of my recommended sodium intake for an entire day. One can of Campbell's "Heart Healthy" Tomato Soup has 42%. One Angus Bacon Cheeseburger from McDonalds has 86%. Grocery shopping takes a lot more time now, because we have to scrutinize every label.

4

Number of times I heard the "d" word come out of a doctor's mouth during my recovery. As in, "You were dead."

13

Number of tests and exams on May 23 and May 24 at the Methodist ER that came back as "normal" before I was cleared for discharge, then suddenly went into ventricular fibrillation without a pulse.

$572,762.80

Total charges for just my hospital stay in the Cardiac ICU at the University of Minnesota. I was in the Cardiac ICU for 26 nights. This does not include professional services such as surgeons, specialists, or doctor consults and visits.

$74,192.82

The total charges for just my hospital stay when I had my ICD implant surgery. My surgery was performed early in the afternoon, and I was kept overnight for observation. This does not include professional services, such as the surgeon's time or doctor consults.

15.5

Number of years it would take to pay off all the accrued hospital and service charges for the entire duration of my care across several different facilities. This does not include professional services, such as specialists. This assumes no interest rate and $5,000 payment each month. Yes, that means my total recovery cost roughly $1 million. As much as it pains me to say it, thank God for insurance companies.

30

Approximately how many minutes your kidneys can go without oxygen before they risk suffering Acute Renal Failure.

4-6

Based on my research, approximately the number of minutes the human brain can survive without sufficient oxygen. Then brain cells start dying off. Like heart cells, brain cells generally do not regenerate.

120+

Number of minutes I was in "prolonged hypoxia" when the doctors evaluated me to see if it was worth trying to initiate ECMO. My body didn't have sufficient oxygen *for more than two hours*.

10

Percentage success rate for adult males recovering from ECMO treatment. Partial recovery, or recovery with significant disability or dysfunction, is higher.

8

Number of days I was on the ECMO. It was surgically initiated on 5/24 and surgically removed 5/31.

61

Number of days from the time I was admitted into the ER until I went home with my family from the Courage Center.

610,000

Number of days it actually felt like between admittance and discharge. Rounded down.

34

Number of facility-administered medications I was given the day my ECMO surgery was initiated.

948

Estimated total number of miles Trish drove back and forth between work/home to visit me at each of the different facilities, from my initial hospitalization through discharge from the Courage Center.

52

Number of distinct delusions I remembered before narrowing it down to the 40-some I included in this book. There are even more kicking around in there, but it is exhausting thinking about all of them.

10

Approximate number of times my meal at the Courage Center would consist of one Styrofoam bowl of mixed salad greens, two cucumber slices, two cherry tomatoes, and fat free dressing. Sometimes I'd get out of control and take three cucumber slices.

0

Number of times the actual phrase "heart attack" appeared in all the medical records I reviewed, as far as I can tell. They've got fancy acronyms and medical terms for it instead – systolic heart failure, cardiogenic shock, non-ST elevation MI, ischemic cardiomyopathy, ventricular fibrillation arrest, and tachyarrhythmia. While, yes, technically they are different physiological events, they are all synonyms for "Not Good."

3

The number of times "stroke" appeared while researching my medical records. They have so many synonyms for heart attack I was stunned when I saw the word stroke, just like us laymen use. I wish there would have been some synonym for that. A stroke just brings to mind an entirely different set of symptoms and repercussions.

25, 27, 33

The holy Trinity of entertainment. These were the channel numbers for ESPN, Fox Sports Midwest, and the Food Network, the three most watched channels I relied on during my stay at the Courage Center.

37

My age when I suffered my first heart attack.

38

My age when I suffered my second heart attack and related complications.

∞

Number of times I heard "You're too young for this."

20. Are We There Yet?

Through the months following my hospitalization, I've kept wondering what exactly it is that I'm doing with my life. Do I pick myself up, dust myself off, and move forward, pretending nothing happened? Has my purpose in life completely changed? What are my goals now? Are they the same they were before all this took place? Should they be? What should I be focusing my time and energy on?

All of these questions have to be addressed in some manner or other. And I guess the process of trying to answer these sorts of questions is part of what it means to be in recovery. I can't help but wonder how I'll know when I'm done. When do I get out of re-covery and move into post-covery? Will I ever get all my covery back? What the heck does recovery even mean at this point? I've heard the word "recovery" so many times that it's started to lose its meaning. Now it sounds like a nonsense word.

Figuring out what my recovery is turns out to be a very interesting question. And a far more complex question than it may first appear. Recovery is a very personal, individualized process. Since no two people and no two triggering events are the same, each person's recovery is necessarily unique. The term recovery is used broadly to define a wide range post-event routines and processes. It is used to refer to physical well-being, mental acuity, spiritual faith and emotional stability. Sometimes individually. Sometimes all at once. It gets very confusing very easily.

Then there's recovery from different perspectives. It can't be solely focused on me, myself and I. The entire family – and to a lesser extent my expanded social network – is also going through a recovery process. What happened to me affected a lot of other people. And what continues to happen to me continues to have ripple effects. Like it or not, for a long time, life was very Jay-centric for everyone around me.

The more time I spent thinking about recovery and what that means to me, the more obsessed I became with defining it. Mapping it out. I wanted to know what the last step in the process was. I wanted to know when I'd be done. Ha. Boy did I have it wrong.

It took a lot of personal reflection and the fellowship of a twelve step program to realize that I was looking at recovery all wrong. It wasn't just my perspective that was wrong. It was my whole concept of what recovery was all about. Things started to go more smoothly once I finally realized that recovery was a *journey* and not a *destination*.

Recovery is a Journey

Accepting that recovery was a journey marked a turning point. Or rather, it marked the point at which my journey finally began. Instead of being fixated on a distant end-point, I could focus on all the things that take place along the way. I could look at the here and now rather than be preoccupied with what *might* be ahead. I could accept that the journey was not a mad dash to the finish line. It was not one huge jump but a series of steps. Sometimes, very small steps.

One of the encouraging things about this perspective is the sense of achievement and progress it allows. There are a number of milestones along the path to recovery. Some days I may make great strides, other days I might struggle and barely inch forward. Regardless, as long as I'm moving forward, I'm making progress. As long as I'm making progress, I'm still on my journey. And as long as I'm still on my journey, I'm recovering.

Recovery as a journey also helps soften the blow of setbacks. It is easy to get discouraged when things don't go as planned, or when I slip back into old habits. Rather than letting these detours become dead ends, I find it's slightly easier to get things back on track with this mindset. As long as I can pull a U-turn and get back on my original course, I'm still moving. I only lose momentum if I come to a complete stop. Otherwise, I'm making progress.

Finding Equilibrium

There are lots of peaks and valleys along the journey. Lots of zig-zagging roads. Lots of obstacles. Lots of chaos. Part of recovery is learning how to apply the right tools to help manage these hazards. Tools that can help create a roadmap of sorts, allowing me to more easily navigate the twists and turns ahead. The further along my journey I go, the better I get at using these tools. And the smoother the road becomes. Or maybe I'm becoming a better driver.

A big part of my recovery is trying to find equilibrium along the way. To slowly but surely bring things back into balance. The Event sort of discombobulated everything. There are a number of aspects of life that need to be rebalanced. Some of these things are severely out of balance, while other things just need a little nudge. Here are some of the things in my life that I'm working on to establish a sense of equilibrium.

- **Centering Focus**. This is probably the biggest thing that's out of balance at the moment. I've been so used to life revolving around me for so long, it's hard to give all that attention up. It's hard to ask for help that was once given freely. It's hard to see things from someone else's point of view when the only view that mattered was mine. It's hard to focus outward when I've spent so much time focused inward.

- **Returning to Routines**. Even simple routines help establish consistency and make the journey more manageable and enjoyable. From taking care of my own personal hygiene again to establishing a regular exercise regimen, or even getting back to work and socializing with co-workers, these routines make it easier to move forward one day at a time.

- **Forming Good Habits**. It can be intimidating to take a serious self-inventory and acknowledge bad habits. Even doing that, it can seem daunting to actually do something about them. But with a slightly different take on things, I've found some success. Rather than focusing on my bad habits, I've tried focusing on what good habits I want to develop. By trying to practice these good habits, I find that some of those bad habits work themselves out.

- **Accepting not Agonizing**. I have spent a considerable amount of time looking backward since my heart attack. It felt like I was always looking over my shoulder, revisiting the horrors I experienced or worrying about symptoms coming back. Well, I've found that it's awfully hard to navigate when you're not facing forward. I can't change what already happened, but I don't need to dwell on it, either.

21. Odds & Ends

There are a number of other strange or noteworthy things that I recall only vaguely. Some of these events don't really fit neatly into any of the categories scattered throughout the other chapters. So I made a special appendix just for them. This is sort of a cathartic, stream of consciousness brain dump for everything else rattling around in my head about The Event. I'm sure I mentioned some of this elsewhere, but hey, apparently it came up again.

- It is an incredibly eerie feeling to return somewhere people said you have been before but you can't remember (my room in the ICU) and be greeted by name by people who recognize you that don't look the least bit familiar (like the nursing staff at the U). And of course, everyone remembered Trish, too. That wasn't quite as eerie.

- When you are trying to remember the details of something that wasn't true in the first place (like my delusion about the Boil Order) it's hard to know if what you're remembering is accurate, since it was all fictional in the first place.

- After a while, you get over modesty. Unless a very pretty young nurse is in the room.

- Yes, it is harder to pee when the nurse stays in the room.

- My brother was working for ESPN at the time of my heart attack. When I first went into the hospital, he put a banner up in the hall for staff to sign and encourage me. Sage Steele, Steve Young, Adam Schefter, and dozens of other people signed the banner for me. My brother even got a photo of Tedy Bruschi signing the banner and wishing me well. The banner is proudly displayed in my mancave.

- Speaking of my brother, we hadn't seen each other in a long time. In fact, we haven't kept in touch very often over the years. I didn't find out until a week or so after regaining consciousness, but he flew in from Bristol on Wednesday, the day after finding out about my heart attack. He stayed until the weekend. Unfortunately, I was a little preoccupied at the time, and missed his entire visit.

- The notes on my 12/31/2011 emergency room visit list my Neurologic state upon admission as "grossly normal" which I found quite amusing. It sounds like an oxymoron, like "a little pregnant."

- When a memory unlocks, it's like being struck by lightning – zap! Suddenly, in a flash, something emerges or crystallizes that wasn't there a split second earlier. It can be hard to imagine it wasn't there the whole time.

- I'm a firm believer in being careful what you wish for. If this ordeal taught me anything, it's that there are some questions you can never un-ask, some facts you can never un-know, and some images you can never un-see. You have to really be sure you can handle the answer before asking a medical question.

- One of the greatest fears I had during my entire hospitalization was of falling asleep. I was afraid that if I fell asleep, I wouldn't wake back up. I kept startling myself awake whenever I would start to drift off. Even now, I cannot fall asleep without medication.

- No marketing material should ever start out with the sentence *"This device was recommended for your use because you are at high risk of sudden cardiac death."* That's the opening line from the LifeVest brochure. I kid you not. Somebody over there needs to get fired.

- It is very, very hard to type on a laptop computer when you have an O2 sat monitor on one finger, you can barely feel your arms, and you're hooked up to various IVs and EKG leads.

- Trish said she was in another room with a social worker and the chaplain when I went into full cardiac arrest, and she could hear me growl and scream in pain each time I was shocked while they tried to resuscitate me. That sent a shiver up my spine.

- Being told things you thought were true for weeks and weeks are all false makes you question everything else for a while, and you're in a constant state of paranoia wondering if anything that isn't occurring at that very moment is real or not.

- Percocet is a potent drug. It knocked out my pain symptoms better than anything else, but also clouded my mind in a tantalizing way. I stopped taking it, worried I would start craving the buzz.

- For my swallow test at Regency, the speech pathologist put blue dye in the taste-test foods. I still had the trach stoma, so they suctioned after the test to see if I had swallowed properly. No blue dye in the vacuum tube meant I was good to go.

- The day I was transferred from the ICU to Regency hospital, there was a man in the next room who went into cardiac arrest and died. I don't know how old he was. But all I could think of for the rest of the day was – *why did he die and I didn't?* I was so distressed by this it made me sick to my stomach.

- One evening there was a wild turkey on the roof, just outside my window. It made me hungry.

- You really find out who you can rely on and trust during tragedy. People step up. And not always the people you expect. But any support, any gesture, no matter how small it may seem, helped. We couldn't have done it without all their help.

- I still can't quite process everything that happened in that span between September 2010 and fall 2011. Two heart attacks. A stroke. Comatose. Paralysis. Induced hypothermia. Pneumonia. Renal failure. Sepsis. Getting shocked back. I don't know how the mind stays sane after all that. Nobody comes equipped to manage that sort of trauma.

- For the first few weeks, nurses or aids had to come in every few hours to shift and rotate my body. Since I was prone in bed for so long, they had to make sure blood didn't pool in parts of my body, so they kept rotating me, rocking me to one side or the other, and wedging pillows under me to help me remain in that new position.

- Mrs. Dash can do wonders for just about any food. Which is important when you're on a low sodium diet. I had Trish buy a few shakers and sneak them in to me. I went through an entire shaker of Mrs. Dash in my 10 days at the Courage Center alone.

- After being in nothing but a hospital gown for more than six weeks, it was bizarre putting regular clothes on. My underwear felt especially strange. I had just started getting used to "ultra-commando" mode.

- When you have short-term memory issues, reading a big book is not the best idea. I would get to a new chapter and suddenly realize I had no idea what happened in the chapter I just finished reading! So then I'd have to go back and start reading it again. Lather, rinse, repeat.

- It got so hot and humid during part of my stay, I remember water condensing on the inside of the windows and slowly dripping down the windowpane. Even that would make me thirsty.

- Knowing you have memory loss is kind of an oxymoron.

- My medical records from Fairview didn't mention anything remarkable about my mental state or coherence. However, my records from Regency reference being in a state of delirium several times – likely caused by the extreme hypoxia and recurring fevers.

- One of the first questions I asked when I came out of sedation completely confused everyone. I asked *"Did Blood Bowl make it to the printers in time?"* No one had any idea what I was talking about. I was referring to a card game called *Blood Bowl Team Manager: The Card Game*, the project I had been working on right before my heart attack. Despite everything going on, I was concerned I may not have delivered all the files needed for the game to get printed in time for a big upcoming convention. Luckily, I had, and it's been one of the most successful projects I've ever worked on.

- Speaking of *Blood Bowl Team Manager*. It debuted at GenCon that August. GenCon is one of the largest gaming conventions in the world, held annually in Indianapolis. I've attended dozens of times, and this was the first GenCon in probably ten years I had to miss. I was emotionally crushed. But FFG surprised me with a wonderful gift – they had everyone from the company and all the visitors who demoed my game sign a copy of the rulebook for me. Every page is covered with signatures and well-wishing. It's one of my most treasured possessions, made even better because people attending the show really liked the game!

- Coban is my friend. It's a special tape made by 3M that only sticks to itself. Most other medical tapes stick very well to skin. Too well. Well enough to completely man-scape my chest and arms.

- That hazy state between sedation and full consciousness is extremely vulnerable. I was pliable and mentally gullible; my brain would have believed anything. Someone could have told me I was recuperating on the moon and I would have believed it. *The moon, you say? Sure, that sounds legit.*

- During a hospitalization after my first heart attack, the blood pressure cuff slipped down from my bicep to my elbow. When it went off, the cuff crushed the IV rig and needles into my arm. Not only was it extremely painful, I also got a huge, nasty looking purple hematoma covering the entire inside of my arm. It lasted for several weeks. I was constantly nervous it would happen again during my second hospitalization, especially with IVs in both arms.

- Unlike most described near-death experiences, I never saw myself as if floating above my own body. I never saw a tunnel of bright light. I never felt inexorably drawn to some other place. I never saw my life flash before my eyes. I was a little bit disappointed.

- Rather than the above, what I saw were spirals. Vibrant swirls and patterns of spiraling specks and shifting colored orbs. A few weeks after my release, I stumbled across a series of amazing photos of galaxy spirals and nebula taken by NASA with their deep space satellites. Those were the spirals I had kept seeing – not something on a quasi-spiritual scale, but something inherently cosmic instead.

- The power source in my ICD implant is rather amazing. It has enough power to potentially deliver more than 100 "treatments" (their euphemism for defibrillation shocks) and still have juice for five to seven years. I can't even get two hours from my cell phone.

- I lost my favorite earring during all the commotion. It's listed on my admittance notes from the ER – it was a small green six-sided die. Considering I could have lost much, much more, it's not such a bad trade-off. But man I loved that earring.

- One of the comments from a chest x-ray taken 6/8/11 reads "Swan-Ganz catheter with tip appreciated in the right pulmonary artery." My initial reaction was "Hey, if that thing was helping me, I'm sure *all* my arteries appreciated it."

- I am incredibly fortunate to not have sustained any significant brain damage, considering how long I was hypoxic. While still sedated, before the doctors could discuss the next course of action, they had to bring me out of hypothermia and "see whether the patient wakes up and is neurologically intact." In other words, they had to check to see if I was already brain dead and beyond help.

- Despite the general consensus that hospital food is terrible, the food at Regency Hospital was actually pretty good. There was a lot of variety, even on a cardiac restricted diet. Granted, I don't know how much of this was actually enjoying the food – or enjoying the fact I could finally eat again.

- Having the occasional Family Movie Night helped things feel normal toward the end of my recovery. I remember watching *The Incredibles*, *Monsters vs Aliens*, *Bolt* and *GalaxyQuest*. Good times.

- Watching the video of my angioplasty procedures is bizarre. You can see them injecting dye as it surges through the network of blood vessels. And you can see right where it stops dead in its tracks. They also have images that clearly show the catheters and stents in place, and how dramatically my blood flow improved after the procedures.

- It was so hot and humid one day at the Courage Center that the two sets of sliding doors at the entrance were dripping with condensation all day long. Every time the doors opened, it felt like a sauna as a wall of hot, sticky heat rolled through the front of the building.

- Managing all the different medications proved quite challenging. In the space of just a few weeks, I had doctor visits and appointments at several different clinics. Each place had a different "current medication list." Some were severely outdated. In one case, the doctor prescribed an increase to one drug, based on the information on one of the outdated lists. Thankfully, Trish caught the error in time – otherwise, the results of taking that medication with my other prescriptions could have been disastrous.

- Thank God I don't have a fear of needles. With the hundreds of shots and blood draws, I would have been a total wreck. There is also a huge variance in quality and precision of the phlebotomists who draw blood. The good ones are in and out before you realize it. The bad ones poke around and wiggle the needle, trying to find a vein.

22. The Big Finish

So that's my story. It's a long one, even without all of the additional anecdotes and side-tracking. Hopefully over the course of the book I was able to both educate and entertain in some fashion. To better explain what actually happened. Not just the heart attack itself, but the impact that The Event – and all the subsequent events it set into motion – had on our lives. The impact it still has on our lives.

So what did I learn from all of this? What did I take away from all these events? Stripping away all of the anecdotes and medical jargon, what impression did this leave on my life? Despite my difficulty in being able to mentally fathom the level of my trauma, I gained *perspective*.

I have a much greater appreciation for the amazing advances in medical technology and training available in this day and age. Now I realize what a powerful impact the nurses and other medical staff have on their patients and how hard they work to provide care. I marvel at the doctors' ability to sift through such vast amounts of medical knowledge – evaluating medications, procedures, surgical options, and other avenues of care to make sure the best course of action is chosen.

I am more grateful than ever for living in the United States and the opportunities it provides, such as access to these medical resources. I feel confident in saying there are very few other places in the world where I could have received the same quality of care and survived both the heart attack and subsequent complications.

I am more certain than ever before in the presence of God and the power of prayer. I have always believed in God, in some manner. But for much of my life, God felt like a more distant, intangible influence. Now I feel like a part of my mind has opened up, allowing me to see more clearly just how present God is in my life, and how truly miraculous my recovery has been. I'm firmly convinced I would not have survived without God's presence and everyone's prayers.

I am humbled and overwhelmed by how remote my chances for survival were. It is staggering, mind-boggling, and even a little bit frightening to really take stock of everything that happened. And everything that had to happen *just right* in order for my recovery to have the outcome it did.

I am also curious as to what survival means from here on out. I find myself spending a lot of time thinking about not just how I survived, but *why* I survived. Is there some deeper meaning I'm supposed to unravel? Some unfulfilled destiny awaiting me? But even less philosophically, there are questions about my capabilities and limitations. Will I be able to continue working? Will I ever recover complete use and feeling in my leg and fingertips? Will I ever be able to sleep normally? Will the nightmares and flashbacks ever end?

Most important of all, I am now hopeful. Hope was a scarce commodity in 2011. Slowly but surely, it's on the rise. With my physical rehabilitation, medications and lifestyle changes there is hope for an improved and fulfilling life from this point on – no matter how long that may be. With continued counseling and therapy there is hope that I can one day exorcise the demons haunting me from the entire ordeal. With everyone's continued support and prayers, there is hope that the love I share with my family is enough to overcome the scars and the scares, and bring us closer together than ever before.

Individually, these may seem like small things compared to everything we went through. Even so, each one is a lens. When I look through all of these lenses collectively, they help snap life into sharper focus, granting me a perspective I did not have before.

I could go on and on. I usually do. But overall, you get the idea. Life has a strange way of coming full circle. So I guess it shouldn't be too surprising that this all ends just how it all began.

With everything being just fine.

(Hopefully Not Quite) The End

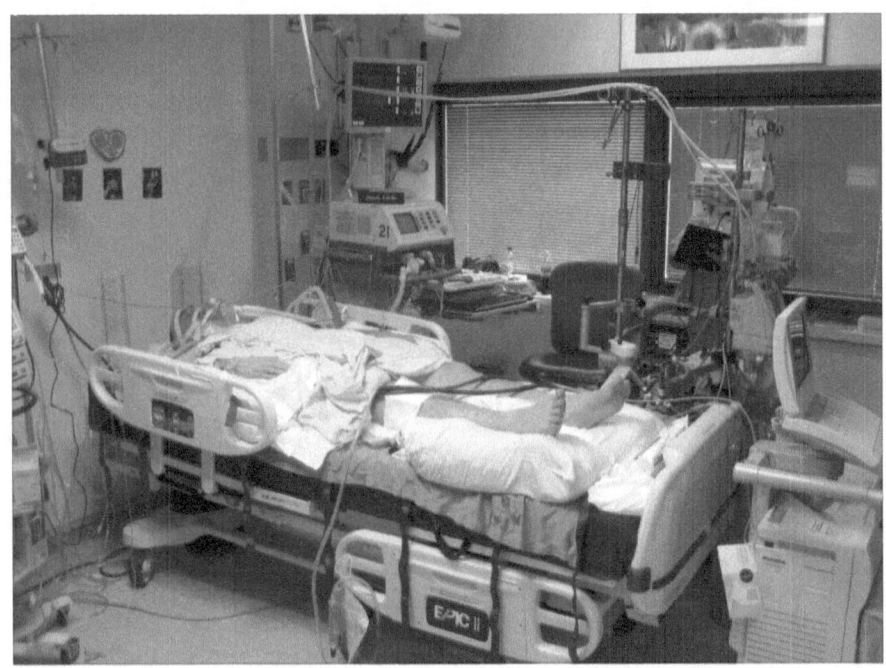

Jay: A picture from the first week in the University of Minnesota's Cardiac ICU. Trish can't remember exactly when this picture was taken, but it was at least a day after my transfer. I'm still hooked up to an aortal balloon pump, ECMO, and respiratory pump. A number of machines had already been cleared out, though. It's hard to imagine more machines crammed in there, with half a dozen people hovering around me checking things. I was completely under during this time, and don't remember this room at all, save for vague memories of seeing webs of tubing and wires overhead.

Trish: And the nurses told us Jay was in one of the smallest rooms in the Cardiac ICU. I'm just glad everything fit.

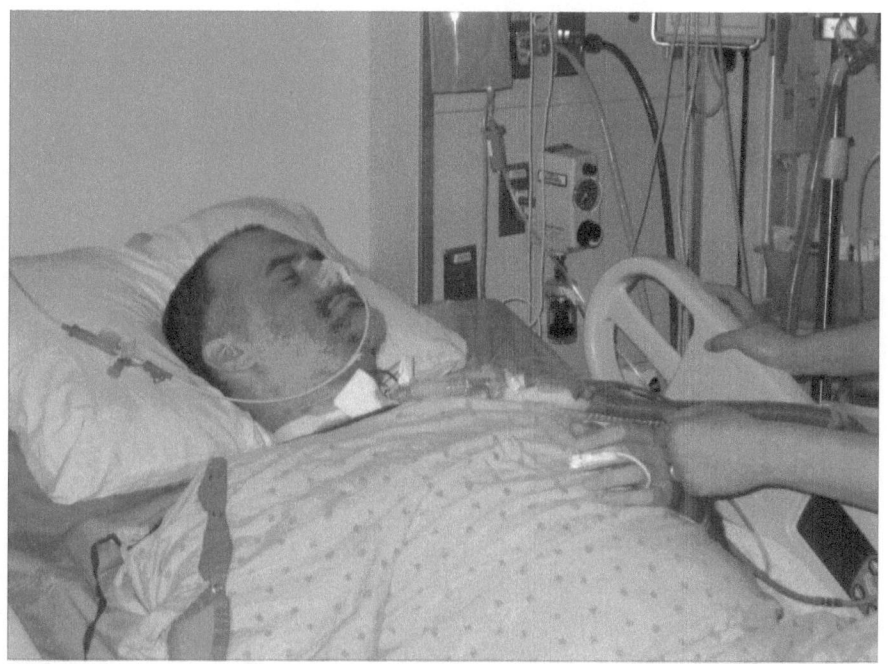

Jay: My trach tube is still in, and I still have the feeding tube – yes, it's inserted through my nose and strung all the way down into my stomach. It was the most disgusting sensation having it removed. It felt like a three foot long, slimy chocolate booger being yanked out. My beard and hair are still pretty short, so this was probably soon after my transfer out of the Cardiac ICU.

Trish: This is actually still in the Cardiac ICU at U of M. But close to his transfer out, due to the trach. While the breathing tube was going through his mouth, there was a big plastic brace holding it on his face; while the trach was scary, at least I got to see his face again.

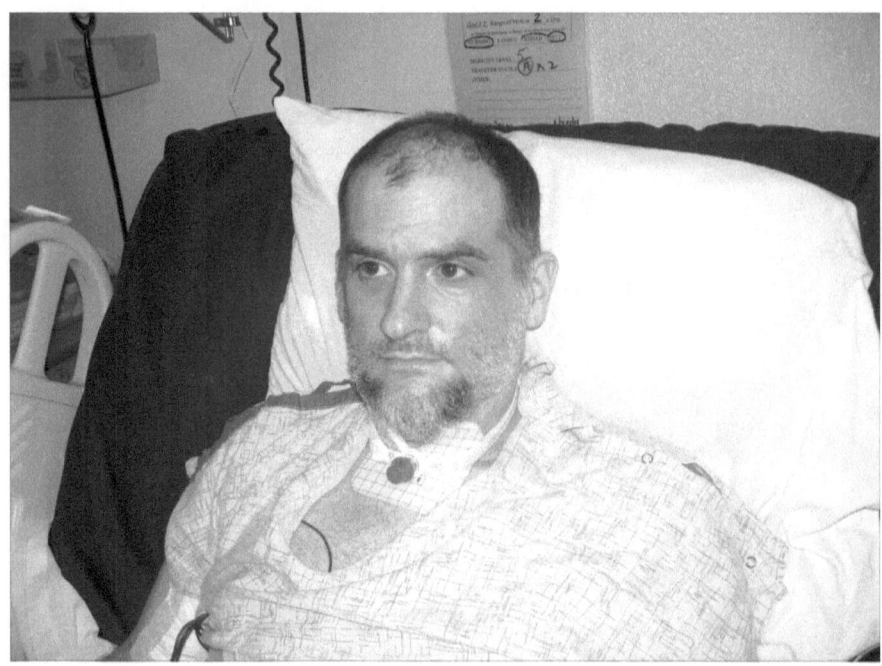

Jay: I look like death warmed over. This must have been very soon after they pulled my trach tube, because the large checkered bandage was only used for a few days. My face is gaunt, and stretched. In the color version, my face is quite yellow, especially around my eyes. This was also after my hair and beard were trimmed at least once. I was pretty clean shaven for a couple of days, and then you can see I scruffed up quite a bit.

Trish: This is from soon after they switched the trach tube to the smaller size that would allow speech. And the feeding tube is gone. This was taken at Regency, and the paper posted to the wall behind his bed gave instructions to the caregivers about what kind of assistance Jay needed to get up and around; beyond that, I don't remember specifics.

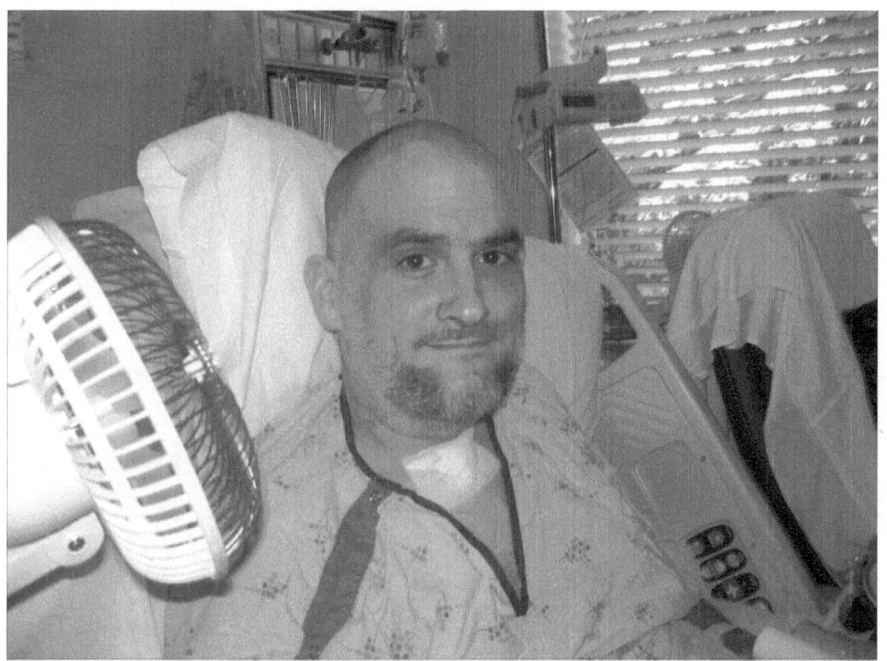

Jay: By stark contrast compared to the previous photos, this picture is from a few days or maybe weeks after my trach had been pulled – you'll notice a much, much smaller dressing covering the site. It's also apparently right after a haircut. I look less gaunt and skeletal than in the earlier photos, and at least there's the look of recognition in my eyes, rather than the vacant, blank stare I usually wore those first few weeks.

Trish: Eating real food was good for Jay's recovery. This was in his last room at Regency, if you look out that window in the background you can see the roof over another wing of the complex. The roof that the wild turkey was pacing along.

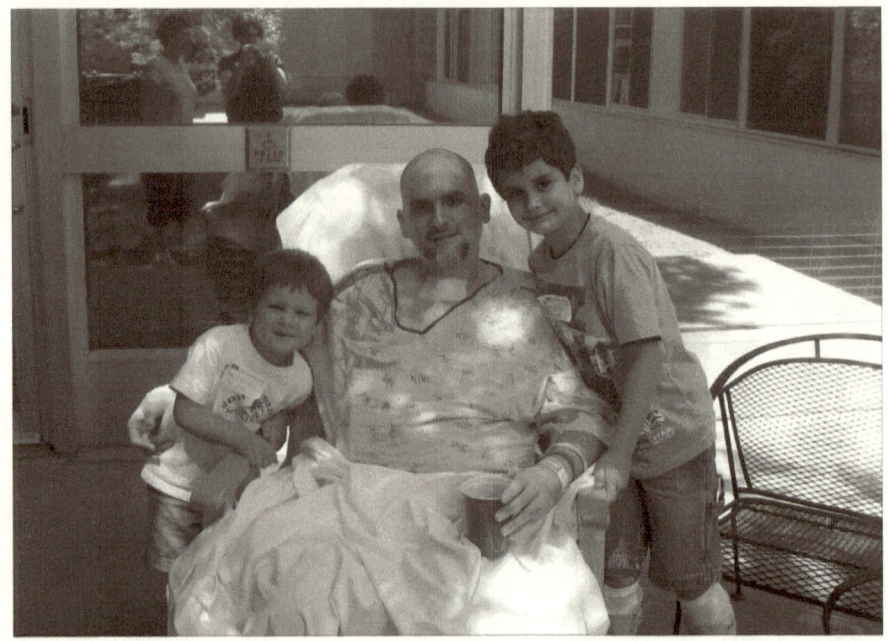

Jay: One of the best days during my recovery. My family came to visit, and I was able to actually get outside. We shared some lemonade. It was wonderful to be surrounded by my family and breathing fresh air. It was such a beautiful day – perfect weather for my first non-ambulance excursion outdoors.

Trish: One of Jay's first wishes he wrote down (that I could read, at least) was to have lemonade with the boys. I'm so glad it worked out, and on such a lovely day.

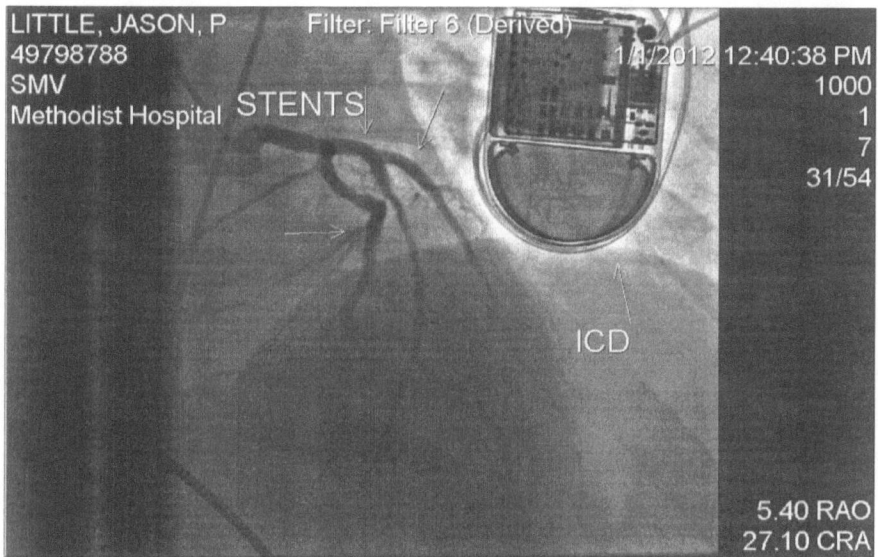

Check it out. Yeah, I know you're jealous. They totally pimped out my heart. New shock absorbers, new motor. I got "The Works" package. This image is actually part of my echocardiogram taken over the New Year's Day weekend from my end-of-2011 emergency room visit. This shows three of the four stents I have, as well as my ICD.

The ICD is about the size and shape of a flip-top metal lighter. It's approximately 75% the size of a deck of playing cards. My left pectoral has finally adjusted to having a chunk of metal between the layer of muscle and my ribcage. The device also has several fine leads threaded into different blood vessels, both to help secure the ICD in place and to take constant readings on my cardiac activity.

It records every heartbeat, every day, and transmits the data electronically to technicians monitoring my device. They alert my cardiologist if there are any irregularities. If I go into complete cardiac arrest, the device will deliver a shock equivalent to the hospital grade defibrillator, enough to hopefully get my heart started again or keep me going long enough for medical personnel to arrive on the scene.

The muscle has grown so snugly around the implant that you can see the entire outline of the top edge of the device. The upper right corner juts out at such a sharp angle, it looks like the ICD is about to wear right through my skin and pop out. Sometimes it feels that way, too. It's positioned three fingers down from my clavicle. That puts it precisely in the way of seatbelts, jumping cats, and overzealous hugs from young children.

Additional Thanks

Here is a list of all the doctors, nurse practitioners, specialists, and other medical staff I found while researching my medical records. I only recognize a few of the names. I'm sure many of these people were helping me survive those first few weeks, while I was sedated. Sorry to all of those I missed. And this doesn't include the nursing staff – another vital part of the process. No matter their level of involvement, each and every person on this list played a part in my recovery.

Thank you to Tadashi Allen, Dr. James Anderson, Asghar, Dr. Leslie Baken, Carol Barsness, Dr. Saba Beg, Dr. Jason Bydash, Nicole Carlson, Pat Christie, Dr. Adina Cioc, Dr. R. Paul Cory, S. Crisman, Dr. Davis, Charles Dietz, Dr. Joseph Dolan, Dr. Michael Dukinfield, Dr. Emmy Earp, Dr. Peter Eckman, Dr. Julie Farias, Dr. Tajudeen Fawole, Brandy Flaten, Jerry Froelich, Dr. Rufino Festin, Dr. Gullickson, Dr. Hagen, Dr. Hillary Hamel, Dr. Tanya Henke-Le, Markus Henning, Dr. Tina C. Huang, Dr. David Ingbar, Dr. Ranjit John, Dr. Aimee Johnson, Louis Kazaglis, Dr. Steven Kind, Dr. Susan Kline, Dr. Christopher Komanapalli, Dr. Suma Konety, Dr. Michele LeClaire, Dr. Kenneth Liao, Jessica Kuehn-Hajder, Dr. Rachel Long, Dr. Fei Lu, Dr. K.P. Madhu, Dr. Richard Madlon-Kay, Richard Mandt, Dr. Lynn Manning, Dr. Cindy Martin, Dr. Sofia Masri, Jason A. Meyers, Dr. Robert McKenna, Dr. Elizabeth Miller, Dr. Emil Missov, Thomas Mullen, LeRoy Munger, Niyada Naksuk, Cathy OBrien, Dr. Mark Pilot, Dr. David Pollak, Dr. Stefan Pambuccian, Dr. Noah Parker, Dr. C. Ross, Dominic Rossini, Cori Russell, Mohammad Sarraf, Dr. Ramiro Saavedra-Romero, Dr. Kent Schwitzer, Dr. Merfake Semret, Dr. Sara Shumway, Dr. Gary Starr, Dr. Theis, Dr. Hemant Trehan, Shashank Vats, Sidney Walker, Neil Wasserman, Dr. Marc Weber, Dr. Katie Willihnganz, Kosuke Yasukawa, and Dr. Gholam Zadeii.

And thank you to my Primary Care Physician, Dr. Aaron Timmerman

About the Author

Born somewhere in California as Jason Patrick Little, the author prefers to go by Jay. He has been an avid gamer for as long as he can remember, and has fond memories of playing Dungeons & Dragons, cribbage, Scrabble, Boggle and countless other games growing up. Board games, card games, roleplaying games, and computer games have always been and will always be a big part of his life.

Over the years, Jay has had the opportunity to work with a number of awesome people in the gaming industry, and has more than thirty nerd-worthy publications to his credit. He has had the chance to work on some amazing licenses and titles like Major League Baseball, Warhammer, and Star Wars. He even got to design a cool game about X-Wings and TIE Fighters and make the *pewpewpew* sound effect during playtesting. This is his first non-gaming related publication.

Jay is also a self-proclaimed geek and proud of it. Thankfully, his wife supports/enables his geekiness. Jay and his wonderful, incredible, awesometastical wife Trish live with their two sons and their cats Jedi and the Meeple in Minnesota. During the winter, Jay wonders why they ever moved to Minnesota. The other four months of the year, he wonders why they didn't do it sooner.

Many animals were harmed in the making of this book, but then Jay was told they were all delusions and never really existed in the first place. So it doesn't really count. Jay found it awkward to write about himself in the third person, but somehow he managed. Thankfully, Jay's sense of humor seems to have returned.

For more information or to learn more about the author, visit his blog:
www.PaintedThumb.com

www.ingramcontent.com/pod-product-compliance
Lightning Source LLC
Chambersburg PA
CBHW032014170526
45157CB00002B/690